# Other Titles in This Series

146 **L. Ya. Adrianova,** Introduction to linear systems of differential equations, 1995
145 **A. N. Andrianov and V. G. Zhuravlev,** Modular forms and Hecke operators, 1995
144 **O. V. Troshkin,** Nontraditional methods in mathematical hydrodynamics, 1995
143 **V. A. Malyshev and R. A. Minlos,** Linear infinite-particle operators, 1995
142 **N. V. Krylov,** Introduction to the theory of diffusion processes, 1995
141 **A. A. Davydov,** Qualitative theory of control systems, 1994
140 **Aizik I. Volpert, Vitaly A. Volpert, and Vladimir A. Volpert,** Traveling wave solutions of parabolic systems, 1994
139 **I. V. Skrypnik,** Methods for analysis of nonlinear elliptic boundary value problems, 1994
138 **Yu. P. Razmyslov,** Identities of algebras and their representations, 1994
137 **F. I. Karpelevich and A. Ya. Kreinin,** Heavy traffic limits for multiphase queues, 1994
136 **Masayoshi Miyanishi,** Algebraic geometry, 1994
135 **Masaru Takeuchi,** Modern spherical functions, 1994
134 **V. V. Prasolov,** Problems and theorems in linear algebra, 1994
133 **P. I. Naumkin and I. A. Shishmarev,** Nonlinear nonlocal equations in the theory of waves, 1994
132 **Hajime Urakawa,** Calculus of variations and harmonic maps, 1993
131 **V. V. Sharko,** Functions on manifolds: Algebraic and topological aspects, 1993
130 **V. V. Vershinin,** Cobordisms and spectral sequences, 1993
129 **Mitsuo Morimoto,** An introduction to Sato's hyperfunctions, 1993
128 **V. P. Orevkov,** Complexity of proofs and their transformations in axiomatic theories, 1993
127 **F. L. Zak,** Tangents and secants of algebraic varieties, 1993
126 **M. L. Agranovskiĭ,** Invariant function spaces on homogeneous manifolds of Lie groups and applications, 1993
125 **Masayoshi Nagata,** Theory of commutative fields, 1993
124 **Masahisa Adachi,** Embeddings and immersions, 1993
123 **M. A. Akivis and B. A. Rosenfeld,** Élie Cartan (1869–1951), 1993
122 **Zhang Guan-Hou,** Theory of entire and meromorphic functions: Deficient and asymptotic values and singular directions, 1993
121 **I. B. Fesenko and S. V. Vostokov,** Local fields and their extensions: A constructive approach, 1993
120 **Takeyuki Hida and Masuyuki Hitsuda,** Gaussian processes, 1993
119 **M. V. Karasev and V. P. Maslov,** Nonlinear Poisson brackets. Geometry and quantization, 1993
118 **Kenkichi Iwasawa,** Algebraic functions, 1993
117 **Boris Zilber,** Uncountably categorical theories, 1993
116 **G. M. Fel′dman,** Arithmetic of probability distributions, and characterization problems on abelian groups, 1993
115 **Nikolai V. Ivanov,** Subgroups of Teichmüller modular groups, 1992
114 **Seizô Itô,** Diffusion equations, 1992
113 **Michail Zhitomirskiĭ,** Typical singularities of differential 1-forms and Pfaffian equations, 1992
112 **S. A. Lomov,** Introduction to the general theory of singular perturbations, 1992
111 **Simon Gindikin,** Tube domains and the Cauchy problem, 1992
110 **B. V. Shabat,** Introduction to complex analysis Part II. Functions of several variables, 1992
109 **Isao Miyadera,** Nonlinear semigroups, 1992
108 **Takeo Yokonuma,** Tensor spaces and exterior algebra, 1992
107 **B. M. Makarov, M. G. Goluzina, A. A. Lodkin, and A. N. Podkorytov,** Selected problems in real analysis, 1992
106 **G.-C. Wen,** Conformal mappings and boundary value problems, 1992
105 **D. R. Yafaev,** Mathematical scattering theory: General theory, 1992

(*Continued in the back of this publication*)

Translations of
MATHEMATICAL
MONOGRAPHS

Volume 146

# Introduction to Linear Systems of Differential Equations

L. Ya. Adrianova

American Mathematical Society
Providence, Rhode Island

Л. Я. Адрианова

# ВВЕДЕНИЕ В ТЕОРИЮ ЛИНЕЙНЫХ СИСТЕМ ДИФФЕРЕНЦИАЛЬНЫХ УРАВНЕНИЙ

Translated from the Russian by Peter Zhevandrov
Translation edited by Simeon Ivanov

1991 *Mathematics Subject Classification.* Primary 34A30, 34Dxx

ABSTRACT. The structure and behavior of solutions of autonomous and periodic systems of ordinary differential equations is considered. Among the properties considered in the book are reducibility of systems, stability of solutions, estimates of solution growth in terms of the coefficients, the influence of a small exponent to the properties of solutions, and central exponents. A complete proof of the necessary and sufficient conditions for the stability of characteristic exponents of two-dimensional diagonal systems is presented.

The book can be used by researchers and graduate students working in differential equations, control theory, and mechanics.

---

**Library of Congress Cataloging-in-Publication Data**

Adrianova, L. Ȋ̑A.

[Vvedenie v teoriȋ̑u lineĭnykh sistem different͡sial′nykh uravneniĭ. English]

Introduction to linear systems of differential equations / L. Ya. Adrianova; [translated from the Russian by Peter Zhevandrov].

p. cm. — (Translations of mathematical monographs, ISSN 0065-9282; v. 146)

Includes bibliographical references (p. – ) and index.

ISBN 0-8218-0328-X (acid-free paper)

1. Differential equations, Linear. I. Title. II. Series.

QA372.A3313 1995

515′.252—dc20

95-35164
CIP

---



⊗ The paper used in this book is acid-free and falls within the guidelines
established to ensure permanence and durability.

♲ Printed on recycled paper.

10 9 8 7 6 5 4 3 2 1 00 99 98 97 96 95

# Contents

# Preface

This textbook is the outgrowth of a lecture course on the theory of linear systems taught by the author at St. Petersburg University the last several years for fourth year students specializing in differential equations.

Linear differential systems are of interest for mathematicians both *per se* and as a tool for studying nonlinear equations by means of the method of linearization. The theory of such systems is rich in problems and methods for their solution [**23**].

It is impossible to include all the basic results and methods in a one-year special course. Indeed, this was not our aim while writing this book, although the material contained here is larger than that given in the lectures. Our goal is to provide an introduction to the theory of linear systems, to acquaint the reader with the basic notions, terms, and definitions of the theory, to show the interrelations among them, and to present some fundamental results and their proofs, in short, to prepare the reader for the study of more involved and specialized parts of the theory.

First we consider properties of solutions of systems with constant and periodic coefficients; this forms the basis for understanding the subsequent material. Here we pay special attention to the construction of a real basis in the case of real coefficients.

Further, by means of the method of characteristic exponents, we study the structure of the space of solutions of a linear system, investigate the properties of reducibility and almost reducibility, and introduce and consider in detail regular systems.

The next part of the book is devoted to the impact of perturbations of the initial data and of the coefficients on the behavior of the solutions. We study various types of stability, perturbations of the coefficients admissible for them, and give estimates of the growth of solutions. One of the most complicated problems of the theory of linear systems is the study of the impact of small perturbations of the coefficients on the characteristic exponents. In order to acquaint the reader with the basic methods for solving this problem, we dwell on the notions of upper and lower functions, central exponents, and integral separateness of a system. Finding necessary and sufficient conditions for the stability of characteristic exponents is one of the fundamental and technically subtle results of recent years, the completion of which relies on the method of rotations due to Millionshchikov [**28, 29**]. Here we give a proof of this result in the case of a two-dimensional diagonal system; this enables us to sufficiently simplify the problem from the technical point of view, preserving the main ideas of the considerations for the general case.

The classical theory of linear systems assumes that the coefficients are bounded. We indicate results that remain valid even when this requirement is weakened. For regular systems this is done in Chapter III, §7. We have tried to emphasize each, athough rare, possibility to connect the properties of solutions with those of the coefficients of a system (see, e.g., Chapter IV, §6, and Chapter VI). There are many examples that precede and conclude the arguments; this is intended to simplify the study of the book.

Naturally, this book borrows some material from basic monographs on the general theory of linear systems [**9, 19**]. At the same time, a number of results given here can be found only in papers published in specialized journals and are not contained in other monographs on differential equations.

The numbering of formulas, theorems, lemmas, etc., is triple (number of chapter, number of section, number of object within section).

N. A. Izobov, a corresponding member of the Belarus Academy of Sciences, has helped the author a lot during the preliminary stage of the preparation of the special course; the collaboration with him has been of great value for the author. Important remarks concerning the plan of the book were furnished by Professor V. A. Pliss, to whom the author is also grateful for his constant support.

# Principal Notation

$\mathbb{C}^n$ is the $n$-dimensional complex vector space,
$\mathbb{R}^n$ is the $n$-dimensional Euclidean space (we write $\mathbb{R}$ instead of $\mathbb{R}^1$),
$\mathbb{Z}$ is the set of integers,
$\mathbb{Z}_+$ is the set of nonnegative integers,
$\mathbb{R}_+$ is the set of nonnegative reals,
$\mathbb{N}$ is the set of naturals,

$x = \begin{pmatrix} x_1 \\ x_2 \\ \dots \\ x_n \end{pmatrix}$ is a vector,

$x^\top = \begin{pmatrix} x_1 \\ x_2 \\ \dots \\ x_n \end{pmatrix}^\top = (x_1, x_2, \dots, x_n)$,

$\|x\|$ is the norm,
$A = \{a_{ij}\}$, $i = 1, \dots, n$, $j = 1, \dots, n$, is the $n \times n$ matrix with the elements $a_{ij}$,
$\operatorname{Sp} A = \sum_{k=1}^{n} a_{kk}$ is the trace of a matrix,
$A^*$ is the Hermitian adjoint for the matrix $A$,
$\|A\| = \max_{\|x\|=1} \|Ax\|$ is the norm of a matrix $A$,
$A_m$ is an $m \times m$ matrix,
$E$ is the identity $n \times n$ matrix,
$E_m$ is the identity $m \times m$ matrix,

$J_\nu(\alpha) = \begin{pmatrix} \alpha & & & 0 \\ 1 & \ddots & & \\ & \ddots & \ddots & \\ 0 & & 1 & \alpha \end{pmatrix}$ is the $\nu \times \nu$ Jordan block,

$C = \operatorname{diag}[c_1, \dots, c_r] = \begin{pmatrix} c_1 & & 0 \\ & c_2 & \\ 0 & & c_r \end{pmatrix}$ is a block-diagonal matrix,

$A_d = \operatorname{diag}[a_{11}, a_{22}, \dots, a_{nn}]$,

an upper-triangular matrix:

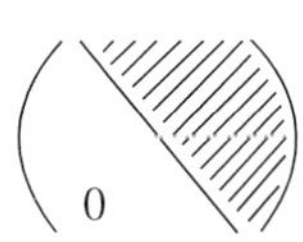

a lower-triangular matrix:

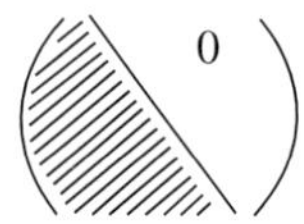

a block-triangular matrix:

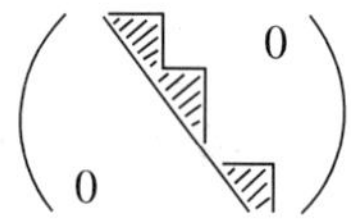

$X(t) = \{x_1(t), \dots, x_n(t)\}$ is the matrix whose columns are vectors $x_1(t), \dots, x_n(t)$,
$X^{-1}(t)$ is the matrix inverse for $X(t)$,
$X(t, \tau) = X(t)X^{-1}(\tau)$,
Det $X$ is the determinant of a matrix $X$,
$A \in C(I)$, $x \in C(I)$ means that the elements of the matrix $A$ and vector $x$ are continuous on the interval $I$,
dim $L$ denotes the dimension of a lineal $L \in \mathbb{C}^n$,
$[t]$ is the integer part of a number $t \in \mathbb{R}$,
$\sigma_X$ is the sum of characteristic exponents of a basis $X$,
$\bar{\lambda} = \overline{\alpha + i\beta} = \alpha - i\beta$ denotes complex conjugation,
$\prec$ is the sign of comparison of growths of functions,
$\chi[\,\cdot\,]$ is the characteristic exponent,
$\delta_{ij}$ is the Kronecker symbol, $\delta_{ij} = \begin{cases} 1, & i = j, \\ 0, & i \neq j, \end{cases}$
$\dot{x} = \dfrac{dx}{dt}$ denotes the derivative.

CHAPTER I

# Linear Autonomous and Periodic Systems

In this and the subsequent chapters we deal with systems of the following form:

$$\dot{x} = A(t)x + f(t), \tag{1.0.1}$$

where $x \in \mathbb{C}^n$, $A(t)$ is a square $n \times n$ matrix, and $f(t)$ is a vector-function with values in $\mathbb{C}^n$. The elements $a_{ij}(t)$ of the matrix $A(t)$ and the coordinates $f_i(t)$ of the vector $f(t)$, $i, j = 1, \dots, n$, are complex functions of the real scalar argument $t$, continuous in some interval $I \subset \mathbb{R}$. In what follows, the latter condition is written as $A \in C(I)$, $f \in C(I)$.

Our goal in this chapter is to recall and somewhat extend the material on linear systems from the general course of differential equations. We consider systems with constant coefficients, the Floquet theory for systems with periodic coefficients, and some results on nonhomogeneous periodic systems.

We begin with recalling the principal terminology.

For any initial data $(t_0, x_0) \in I \times \mathbb{C}^n$, the solution of the Cauchy problem exists, is unique, and is defined for all $t \in I$.

The system (1.0.1) is a *linear nonhomogeneous system*; the system

$$\dot{x} = A(t)x \tag{1.0.2}$$

is *linear homogeneous*. Any set of $n$ linearly independent solutions $x_1(t), \dots, x_n(t)$ of system (1.0.2) is called a *fundamental system of solutions* and is a basis in the space of its solutions. A matrix $X(t) = \{x_1(t), \dots, x_n(t)\}$ whose colums are the vectors of a basis is called a *fundamental matrix*. Obviously, such a matrix is a solution of the matrix equation

$$\frac{dX}{dt} = A(t)X, \tag{1.0.3}$$

and, conversely, any nonsingular solution of equation (1.0.3) is a fundamental matrix of system (1.0.2). If one fundamental matrix $X_1(t)$ is known, then the complete set of fundamental matrices has the form

$$X(t) = X_1(t)C, \tag{1.0.4}$$

where $C$ is a constant nonsingular matrix. From the Ostrogradskiĭ-Jacobi-Liouville formula

$$\operatorname{Det} X(t) = \operatorname{Det} X(t_0) \exp \int_{t_0}^{t} \operatorname{Sp} A(u)\, du, \qquad t, t_0 \in I, \tag{1.0.5}$$

it follows that it is sufficient to verify the nonsingularity of the solution of the matrix equation (1.0.3) at a single point. A fundamental matrix $X(t)$ is said to be *normalized at a point* $t_0 \in I$, if $X(t_0) = E$; then such a matrix is written as $X(t) = X(t, t_0)$.

If $X(t)$ is a fundamental matrix of system (1.0.2), then the general solution of the linear homogeneous system is

$$x = X(t)c, \qquad \text{where} \qquad c \in \mathbb{C}^n,$$

and that of the nonhomogeneous one is

$$x = X(t)c + X(t)\int X^{-1}(t)f(t)\,dt, \qquad t \in I;$$

the latter follows from the method of variation of parameters.

The solution of the Cauchy problem with initial data $(t_0, x_0) \in I \times \mathbb{C}^n$ is written for the homogeneous system as

$$x(t) = X(t)X^{-1}(t_0)x_0 = X(t, t_0)x_0,$$

and for a nonhomogeneous one as

$$\begin{aligned} x(t) &= X(t)X^{-1}(t_0)x_0 + \int_{t_0}^{t} X(t)X^{-1}(\tau)f(\tau)\,d\tau \\ &= X(t, t_0)x_0 + \int_{t_0}^{t} X(t, \tau)f(\tau)\,d\tau. \end{aligned}$$

The matrix $X(t, \tau)$, where $t, \tau \in I$, is called the Cauchy matrix of system (1.0.2).

## §1. Adjoint systems

Together with a system

$$\dot{x} = A(t)x \tag{1.1.1}$$

we shall consider the system

$$\dot{y} = -A^*(t)y, \tag{1.1.2}$$

where $A^*$ is the Hermitian adjoint of $A(t)$.

DEFINITION 1.1.1. The system (1.1.2) is called *adjoint* for system (1.1.1).

Let us study the relations between the solutions of these systems.

THEOREM 1.1.1. *If $X(t)$ is a fundamental matrix of system* (1.1.1), *then*

$$[X^{-1}(t)]^*$$

*is a fundamental matrix of the adjoint system* (1.1.2).

PROOF. The matrix $X(t)$ is nonsingular; hence, $X^{-1}(t)$ exists. Differentiating the identity $X(t)X^{-1}(t) \equiv E$ for $t \in I$,

$$\frac{dX}{dt}X^{-1} + X\frac{dX^{-1}}{dt} \equiv 0,$$

we have

$$\frac{dX^{-1}}{dt} = -X^{-1}\frac{dX}{dt}X^{-1} = -X^{-1}A(t)XX^{-1} = -X^{-1}A(t).$$

The proof is completed by passing to adjoint matrices in the last identity. □

THEOREM 1.1.2. *If $X(t)$ is a fundamental matrix for system* (1.1.1), *then a matrix $Y(t)$ is fundamental for the adjoint system* (1.1.2) *if and only if the identity*

$$Y^*(t)X(t) \equiv C, \qquad t \in I, \tag{1.1.3}$$

*where $C$ is a nondegenerate matrix constant, is satisfied.*

PROOF. Let $Y(t)$ be a fundamental matrix for (1.1.2). Let us show that identity (1.1.3) is valid. According to Theorem 1.1.1, $[X^{-1}(t)]^*$ is also fundamental for (1.1.2); hence, there exists a nondegenerate matrix $C$ such that $Y(t) = [X^{-1}(t)]^*C$. We pass to adjoint matrices and obtain the statement required. To prove the sufficiency we find $Y(t)$ from (1.1.3), i.e., $Y(t) = [X^{-1}(t)]^*C$. According to Theorem 1.1.1, the matrix $Y(t)$ is fundamental for (1.1.2). □

REMARK 1.1.1. In the case $A^*(t) = -A(t)$, $t \in I$, system (1.1.1) is called *selfadjoint*. It follows from identity (1.1.3) that the Euclidean length of each vector-solution of this system is constant.

## §2. The matriciant and its properties

We assume that the norm introduced in the set of $n \times n$ square matrices $A$ is induced by the norm introduced in the set of $n$-dimensional vectors $x$, i.e.,

$$\|A\| = \max_{\|x\|=1} \|Ax\|.$$

It is known **[36]** that any such norm is multiplicative,

$$\|A_1A_2\| \leqslant \|A_1\| \cdot \|A_2\|,$$

and is consistent with the norm of the vectors,

$$\|Ax\| \leqslant \|A\| \cdot \|x\|.$$

In what follows we shall use one of the three forms:

$$\|x\|_I = \max_i |x_i| \quad \Rightarrow \quad \|A\|_I = \max_i \sum_{j=1}^n |a_{ij}|,$$

$$\|x\|_{II} = \sum_{i=1}^n |x_i| \quad \Rightarrow \quad \|A\|_{II} = \max_j \sum_{i=1}^n |a_{ij}|,$$

$$\|x\|_{III} = (x, x)^{1/2} \quad \Rightarrow \quad \|A\|_{III} = (\text{the greatest eigenvalue of } A^*A)^{1/2}.$$

Here $x_i$, $i = 1, \dots, n$, are the elements of the vector $x$, and $a_{ij}$, $i, j = 1, \dots, n$, are the elements of the matrix $A$. All these norms have the properties indicated above. If the choice of the norm in the reasoning is of no principal importance, the indices will be omitted. If the elements of the matrix $A$ and of the vector $x$ depend on $t$, then the norm becomes a function of $t$.

Let us return to the system

$$\dot{x} = A(t)x, \qquad A \in C(I), \quad x \in \mathbb{C}^n, \tag{1.2.1}$$

and try to understand the dependence of the fundamental matrix on the coefficients of the system. To this end, by means of successive approximations we solve the Cauchy matrix problem

$$\frac{dX}{dt} = A(t)X, \qquad X(t_0) = E. \tag{1.2.2}$$

Thus, we have

$$\begin{aligned} &X_0(t) = E, \\ &X_k(t) = E + \int_{t_0}^{t} A(u)X_{k-1}(u)\,du, \qquad k = 1, 2, \dots, \qquad t \in I, \end{aligned} \tag{1.2.3}$$

or

$$\begin{aligned} X_k(t) = E &+ \int_{t_0}^{t} A(u)\,du + \int_{t_0}^{t} A(t_1)\,dt_1 \int_{t_0}^{t_1} A(t_2)\,dt_2 + \cdots \\ &+ \int_{t_0}^{t} A(t_1)\,dt_1 \int_{t_0}^{t_1} A(t_2)\,dt_2 \cdots \int_{t_0}^{t_{k-1}} A(t_k)\,dt_k. \end{aligned} \tag{1.2.4}$$

Let us show that the sequence (1.2.3) converges absolutely for $t \in I$ and that this convergence is uniform in any closed interval. Indeed, the convergence of the sequence (1.2.3) is equivalent to the convergence of the series

$$X_0 + (X_1(t) - X_0) + \cdots + (X_k(t) - X_{k-1}(t)) + \cdots. \tag{1.2.5}$$

By induction, for the terms of this series we have the estimate

$$\|X_m(t) - X_{m-1}(t)\| \leqslant \frac{1}{m!}\left(\left|\int_{t_0}^{t} \|A(u)\|\,du\right|\right)^m, \qquad m = 1, 2, \dots, \qquad t \in I. \tag{1.2.6}$$

For $m = 1$ the estimate is obvious. The passage from $m$ to $m+1$ is carried out by taking into account (1.2.4):

$$\begin{aligned} \|X_{m+1}(t) - X_m(t)\| &= \left\| \int_{t_0}^{t} A(t_1)\,dt_1 \int_{t_0}^{t_1} A(t_2)\,dt_2 \cdots \int_{t_0}^{t_m} A(t_{m+1})\,dt_{m+1} \right\| \\ &\leqslant \left| \int_{t_0}^{t} \|A(t_1)\|\,dt_1 \frac{1}{m!}\left(\left|\int_{t_0}^{t} \|A(u)\|\,du\right|\right)^m \right| \\ &= \frac{1}{(m+1)!}\left(\left|\int_{t_0}^{t} \|A(u)\|\,du\right|\right)^{m+1}. \end{aligned}$$

Thus, the series (1.2.5) is majorized by the convergent series (see (1.2.6))

$$1 + \sum_{k=1}^{\infty} \frac{1}{k!}\left(\left|\int_{t_0}^{t} \|A(u)\|\,du\right|\right)^k = \exp\left|\int_{t_0}^{t} \|A(u)\|\,du\right|; \tag{1.2.7}$$

this implies its convergence with the properties indicated above.

Denoting the sum of series (1.2.5) by $\Omega_{t_0}^{t} A$, we obtain

$$\underset{t_0}{\overset{t}{\Omega}} A = E + \int_{t_0}^{t} A(u)\,du + \sum_{k=2}^{\infty}\left(\int_{t_0}^{t} A(t_1)\,dt_1 \cdots \int_{t_0}^{t_{k-1}} A(t_k)\,dt_k\right). \tag{1.2.8}$$

Differentiating the series (1.2.8) term by term, we obtain the same series multiplied on the left by $A(t)$, and the latter converges uniformly on any closed subinterval of $I$; hence, the identity

$$\frac{d\,\Omega_{t_0}^t A}{dt} = A(t)\,\overset{t}{\underset{t_0}{\Omega}} A$$

is valid. The matrix $\Omega_{t_0}^t A$, i.e., the solution of problem (1.2.2), is called the *matriciant* of system (1.2.1). This is the fundamental matrix normalized at $t = t_0$ and connected with any other fundamental matrix $X(t)$ in the following way:

$$\overset{t}{\underset{t_0}{\Omega}} A = X(t)X^{-1}(t_0) = X(t, t_0).$$

The notation for the matriciant appeals to the system by which it is generated; the series (1.2.8) gives its expression via the coefficients of this system.

Note some properties of the matriciant.

1. $\|\Omega_{t_0}^t A\| \leqslant \exp\left|\int_{t_0}^t \|A(u)\|\,du\right|$ (this follows from (1.2.7)).

2. In the case of a constant matrix $A$ we have

$$\overset{t}{\underset{t_0}{\Omega}} A = \exp A(t - t_0),$$

which immediately follows from (1.2.8), since the series (1.2.8) is nothing else but the expansion of $\exp A(t - t_0)$ **[16, 19]**.

3. The multiplicative property of the matriciant:

$$\overset{t}{\underset{t_0}{\Omega}} A = \overset{t}{\underset{t_1}{\Omega}} A\, \overset{t_1}{\underset{t_0}{\Omega}} A, \qquad t_0, t_1, t \in I.$$

The fundamental matrices on either side of the equality coincide for $t = t_1$; hence, by virtue of uniqueness, they coincide for all $t \in I$.

4. If $A(t_1)$ and $B(t_2)$ comute for any $t_1, t_2 \in I$, then

$$\overset{t}{\underset{t_0}{\Omega}}(A + B) = \overset{t}{\underset{t_0}{\Omega}} A\, \overset{t}{\underset{t_0}{\Omega}} B.$$

Let us denote the left-hand side of the equality by $\Lambda(t)$ and the right-hand side by $\Pi(t)$. By definition, $\dot{\Lambda} = (A(t) + B(t))\Lambda$. Now we write $\dot{\Pi}$:

$$\dot{\Pi} = A(t)\Pi + \overset{t}{\underset{t_0}{\Omega}} A B(t) \overset{t}{\underset{t_0}{\Omega}} B = A(t)\Pi + B(t)\Pi = [A(t) + B(t)]\Pi.$$

Thus, the matrices $\Lambda(t)$ and $\Pi(t)$ satisfy the same matrix differential equation and, moreover, $\Lambda(t_0) = \Pi(t_0) = E$, which, by virtue of uniqueness, implies $\Lambda(t) \equiv \Pi(t)$ for $t \in I$.

REMARK 1.2.1. If constant matrices $A$ and $B$ commute, then

$$e^{A+B} = e^A e^B.$$

Indeed,

$$\exp(A + B) = \overset{t}{\underset{0}{\Omega}}(A + B) = \overset{1}{\underset{0}{\Omega}} A\, \overset{1}{\underset{0}{\Omega}} B = \exp A \exp B.$$

5. The expansion of the matriciant in powers of a parameter. Let the system

$$\dot{x} = A(t, \varepsilon)x \tag{1.2.9}$$

be such that

$$A(t,\varepsilon)=\sum_{k=0}^{\infty}A_k(t)\varepsilon^k, \tag{1.2.10}$$

where the series converges absolutely, and the series of the norms of its terms converges uniformly in $t\in I$, $|\varepsilon|<\mathcal{E}$; moreover, let $A(t,\varepsilon)$, $A_k(t)$ be continuous for $t\in I$. Then the matriciant of system (1.2.9) can be expanded in a series in powers of $\varepsilon$ that converges absolutely for $t\in I$, $|\varepsilon|<\mathcal{E}$, and whose coefficients are continuous for $t\in I$.

The proof is carried out by the method of majorants.

a) The construction of a formal solution. We shall seek the solution of the matrix equation

$$\frac{dX}{dt}=A(t,\varepsilon)X \tag{1.2.11}$$

in the form of a series

$$X(t,\varepsilon)=X_0(t)+X_1(t)\varepsilon+\cdots+X_k(t)\varepsilon^k+\cdots, \tag{1.2.12}$$

such that

$$X(t_0,\varepsilon)=E. \tag{1.2.13}$$

We substitute the series (1.2.12) in equation (1.2.11) and, taking into account expansion (1.2.10), equate the coefficients at like powers of $\varepsilon$. We obtain the following equations:

$$\frac{dX_0}{dt}=A_0(t)X_0,$$

$$\frac{dX_k}{dt}=A_0(t)X_k+\sum_{i=1}^{k}A_i(t)X_{k-i},\qquad k\geqslant 1.$$

Let us choose the initial conditions for the solutions of these equations so that condition (1.2.13) is satisfied independently of the specific value of $\varepsilon$. Obviously, we have to set

$$X_0(t_0)=E,\qquad X_k(t_0)=0,\qquad k\geqslant 1.$$

Thus, we obtain the following formulas for the coefficients of series (1.2.12):

$$X_0(t)=\overset{t}{\underset{t_0}{\Omega}}A_0,$$

$$X_k(t)=\int_{t_0}^{t}\overset{t}{\underset{u}{\Omega}}A\sum_{i=1}^{k}A_i(u)X_{k-i}(u)\,du,\qquad k\geqslant 1.$$

Thus, a formal solution is constructed and it remains to prove its convergence; then the epithet "formal" can be omitted.

b) The proof of convergence of series (1.2.12). The idea of the proof is as follows. Let us construct a scalar equation majorizing the initial system (1.2.9), find its solution normalized at $t=t_0$, and show that this solution defines a series which is majorant for the formal solution (1.2.12). The required result will follow from the convergence of the majorant series.

Let $0 < \varepsilon_1 < \mathcal{E}$. By the absolute convergence of series (1.2.10), we have

$$\sum_{k=0}^{\infty} \|A_k(t)\|\varepsilon_1^k = M(t);$$

thus,

$$\|A_k(t)\| \leqslant \frac{M(t)}{\varepsilon_1^k}, \qquad t \in I. \tag{1.2.14}$$

The last inequality implies that the series (1.2.10) is majorized by the convergent series

$$\sum_{k=0}^{\infty} M(t)\left(\frac{\varepsilon}{\varepsilon_1}\right)^k = M(t)\frac{1}{1-\varepsilon/\varepsilon_1}, \qquad |\varepsilon| < \varepsilon_1, \quad t \in I.$$

The subsequent constructions are carried out for $t \geqslant t_0$. Let us consider the equation

$$\frac{dy}{dt} = M(t)\frac{y}{1-\varepsilon/\varepsilon_1}, \qquad y(t_0) = 1, \tag{1.2.15}$$

and write its solution

$$y(t) = \exp\int_{t_0}^{t} \frac{M(\tau)}{1-\varepsilon/\varepsilon_1}\,d\tau$$

in the form of a series

$$y(t) = y_0(t) + y_1(t)\varepsilon + \cdots + y_k(t)\varepsilon^k + \cdots. \tag{1.2.16}$$

Note that this series converges for $t \geqslant t_0$, $|\varepsilon| < \varepsilon_1$. Let us show that the series (1.2.16) majorizes the series (1.2.12) defining the formal solution, which completes our proof. The coefficients of the series (1.2.16) can be defined by substituting it in equation (1.2.15) and equating the coefficients at like powers of $\varepsilon$ (as we have done in item a)):

$$\frac{dy_0}{dt} = M(t)y_0, \qquad y_0(t_0) = 1.$$

From this, according to (1.2.14), we have

$$\|X_0(t)\| = \left\|\overset{t}{\underset{t_0}{\Omega}} A_0\right\| \leqslant \exp\int_{t_0}^{t} \|A_0(\tau)\|\,d\tau \leqslant \exp\int_{t_0}^{t} M(\tau)\,d\tau = y_0(t).$$

Further,

$$\frac{dy_k}{dt} = M(t)\left[y_k + \frac{y_{k-1}}{\varepsilon_1} + \cdots + \frac{y_0}{\varepsilon_1^k}\right], \qquad y_k(t_0) = 0, \quad k \geqslant 1.$$

This implies

$$y_k(t) = \int_{t_0}^{t} e^{\int_u^t M(\tau)\,d\tau}\left[\frac{M(u)}{\varepsilon_1}y_{k-1}(u) + \cdots + \frac{M(u)}{\varepsilon_1^k}y_0(u)\right]du, \qquad k \geqslant 1.$$

By induction, taking into account (1.2.14), we obtain

$$\|X_k(t)\| \leqslant y_k(t), \qquad t \geqslant t_0, \quad k = 0, 1, 2, \ldots.$$

For $t \leqslant t_0$, the series (1.2.12) is majorized by the solution of the equation

$$\frac{dy}{dt} = -M(t)\frac{y}{1-\varepsilon/\varepsilon_1}, \qquad y(t_0) = 1. \quad \square$$

## §3. Linear systems with constant coefficients

Consider a system

$$\dot{x} = Ax \tag{1.3.1}$$

with a constant matrix $A$, i.e., an autonomous system. In §2 we showed that

$$\overset{t}{\underset{0}{\Omega}} A = X(t,0) = e^{At}.$$

Let us find the structure of this fundamental matrix. Let $S$ be the matrix transforming $A$ to its Jordan canonical form, i.e.,

$$B = S^{-1}AS = \operatorname{diag}[J_{p_1}(\lambda_1), \dots, J_{p_k}(\lambda_k)], \tag{1.3.2}$$

where $k \leqslant n$, $\sum_1^k p_i = n$, $\lambda_i$ are the eigenvalues of the matrix $A$ (some of them can be equal), and $J_\nu(\lambda)$ is a $\nu \times \nu$ Jordan block,

$$J_\nu(\lambda) = \begin{pmatrix} \lambda & 0 & \dots & \dots & 0 \\ 1 & \lambda & \dots & \dots & \vdots \\ 0 & \dots & \ddots & \dots & \vdots \\ \vdots & \dots & \dots & \ddots & \vdots \\ 0 & \dots & 0 & 1 & \lambda \end{pmatrix}.$$

According to the properties of the matrix exponential **[16, 19]** (which are verified by its straightforward expansion in a series), we have

$$e^{At} = e^{SBS^{-1}t} = Se^{Bt}S^{-1} \tag{1.3.3}$$

and

$$e^{Bt} = \operatorname{diag}\left[e^{J_{p_1}(\lambda_1)t}, \dots, e^{J_{p_k}(\lambda_k)t}\right].$$

Let us find out what the diagonal blocks of this matrix are,

$$\begin{aligned} e^{J_\nu(\lambda)t} &= e^{E_\nu \lambda t + J_\nu(0)t} = e^{E_\nu \lambda t} e^{J_\nu(0)t} \\ &= e^{\lambda t} e^{J_\nu(0)t} = e^{\lambda t} \sum_{k=0}^{\infty} \frac{1}{k!} J_\nu^k(0) t^k. \end{aligned}$$

$J_\nu(0)$ is a nilpotent matrix; thus, $\exp J_\nu(0)t$ is a finite sum of $\nu$ terms and can be easily calculated:

$$e^{J_\nu(\lambda)t} = e^{\lambda t} \begin{pmatrix} 1 & 0 & \dots & 0 \\ t & 1 & \dots & \vdots \\ \vdots & & \ddots & 0 \\ \dfrac{t^{\nu-1}}{(\nu-1)!} & \dots & t & 1 \end{pmatrix}. \tag{1.3.4}$$

If the matrix $A$ is complex, then our problem is solved. If the matrix $A$ is real, then $\exp At$ is real, which follows from the construction of the matriciant. If all the eigenvalues of the matrix $A$ are real, then the matrices $S$ and $B$ are real and the expression for $\exp At$ in the form (1.3.3) does not need further explanations. On the other hand, when there are complex eigenvalues of the matrix $A$, then the matrices $S$ and $B$ are complex, and the expression for $\exp At$, the latter being real, does need further clarification.

Let the matrix $A$ be real and let the Jordan block $J_{\rho_j}(\lambda_j)$ of order $v$ correspond to its eigenvalue $\lambda_j = \alpha + i\beta$ in (1.3.2). Since $A$ is real, there exists an eigenvalue $\lambda_{j+1} = \bar{\lambda}_j$ with the corresponding Jordan block

$$J_{\rho_{j+1}}(\lambda_{j+1}) = \overline{J_{\rho_j}(\lambda_j)}.$$

Let us consider the diagonal element of the matrix $B$ of dimension $2v \times 2v$ containing these blocks ($v = \rho_j = \rho_{j+1}$; the index of $\lambda$ is omitted):

$$B_{2v} = \begin{pmatrix} J_v(\alpha + i\beta) & 0 \\ 0 & J_v(\alpha - i\beta) \end{pmatrix} = \begin{pmatrix} J_v(\alpha) + i\beta E_v & 0 \\ 0 & J_v(\alpha) - i\beta E_v \end{pmatrix},$$

and carry out a special similarity transformation of the matrix $B_{2v}$ reducing it to a real form. This standard transformation has the following form:

$$S_{2v} = \begin{pmatrix} E_v & iE_v \\ E_v & -iE_v \end{pmatrix}, \qquad S_{2v}^{-1} = \frac{1}{2}\begin{pmatrix} E_v & E_v \\ -iE_v & iE_v \end{pmatrix}.$$

Thus, we have

$$\widetilde{B}_{2v}(\lambda) = S_{2v}^{-1} B_{2v} S_{2v} = \begin{pmatrix} J_v(\alpha) & -\beta E_v \\ \beta E_v & J_v(\alpha) \end{pmatrix}.$$

The matrix $\widetilde{B}_{2v}$ is called the real canonical form of the matrix $B_{2v}$.

By means of the similarity transformation with the matrix $S$, we reduce the matrix $A$ to the form (1.3.2); then to each pair of complex-conjugate Jordan blocks (there are no other complex blocks) we apply the transformation of the form $S_{2v}$ and finally obtain the real matrix

$$\widetilde{B} = \operatorname{diag}[J_{\rho_1}(\lambda_1), \dots, J_{\rho_m}(\lambda_m), \widetilde{B}_{2v_{m+1}}(\lambda_{m+1}), \dots, \widetilde{B}_{2v_{k-1}}(\lambda_{k-1})],$$

where the first $m$ blocks correspond to real eigenvalues of the matrix $A$ and the subsequent blocks correspond to the complex ones. The matrix $\widetilde{B}$ is similar to the matrix $A$ and is obtained from it by means of a repeated complex transformation. Can we choose this transformation so that it is real?

LEMMA 1.3.1. *If matrices $A$ and $B$ are real and there exists a complex matrix $S$ such that*

$$B = S^{-1}AS,$$

*then there exists a real matrix $S_0$ realizing this equality.*

PROOF. Let $S = S_1 + iS_2$, where $S_1$ and $S_2$ are real. From $AS = SB$ we have $A(S_1 + iS_2) = (S_1 + iS_2)B$ or $AS_1 = S_1 B$, $AS_2 = S_2 B$. The matrices $S_i$, $i = 1, 2$, are real but they may be singular. We continue with the proof. We multiply the last equality by a number $\rho$ and add it to the preceding one:

$$A(S_1 + \rho S_2) = (S_1 + \rho S_2)B.$$

Note that $\operatorname{Det}(S_1 + \rho S_2)$ is a polynomial in $\rho$ of degree at most $n$ with real coefficients not equal simultaneously to zero, since $\operatorname{Det}(S_1 + iS_2) \neq 0$. Hence, there exists a $\rho_0 \in \mathbb{R}$ such that the matrix $S_0 = S_1 + \rho_0 S_2$ is nondegenerate. $\square$

It follows from Lemma 1.3.1 that there exists a real matrix $S_0$ such that $\widetilde{B} = S_0^{-1} A S_0$; therefore, we have

$$
(1.3.5)\qquad \begin{aligned} e^{At} &= S_0 e^{\widetilde{B}t} S_0^{-1} \\ &= S_0 \operatorname{diag}\left[e^{J_{p_1}(\lambda_1)t}, \dots, e^{J_{p_m}(\lambda_m)t}, e^{\widetilde{B}_{2\nu_{m+1}}(\lambda_{m+1})t}, \dots, e^{\widetilde{B}_{2\nu_{k-1}}(\lambda_{k-1})t}\right] S_0^{-1}. \end{aligned}
$$

The form of the first $m$ blocks on the diagonal is given in (1.3.4). Now we consider the other ones:

$$
\begin{aligned} e^{\widetilde{B}_{2\nu}(\lambda)t} &= S_{2\nu}^{-1} e^{B_{2\nu}(\lambda)t} S_{2\nu} = S_{2\nu}^{-1} \begin{pmatrix} e^{J_\nu(\alpha)t} e^{i\beta t} & 0 \\ 0 & e^{J_\nu(\alpha)t} e^{-i\beta t} \end{pmatrix} S_{2\nu} \\ &= \operatorname{diag}\left[e^{J_\nu(\alpha)t}, e^{J_\nu(\alpha)t}\right] \begin{pmatrix} E_\nu \cos\beta t & -E_\nu \sin\beta t \\ E_\nu \sin\beta t & E_\nu \cos\beta t \end{pmatrix}. \end{aligned}
$$

Thus, we have finally found the form of the fundamental matrix $e^{At}$ of system (1.3.1) in relation to the Jordan canonical form of the matrix $A$.

REMARK 1.3.1 (on the behavior of the solutions of system (1.3.1) as $t \to \infty$). The form obtained for the matriciant of system (1.3.5) explicitly demonstrates that the behavior of the solutions as $t$ grows depends on the form of the eigenvalues of the matrix $A$ and its elementary divisors. Indeed, one of the fundamental matrices of system (1.3.1) is the matrix

$$
X(t) = S_0 e^{\widetilde{B}t},
$$

where the first $p_1$ column-solutions are generated by the block $J_{p_1}(\lambda_1)t$, the second $p_2$ column-solutions are generated by the block $\exp J_{p_2}(\lambda_2)t$, and, finally, the last ones by the block $\exp \widetilde{B}_{2\nu_{k-1}}(\lambda_{k-1})t$. This implies the validity of the following statements. Let $\lambda$ be an eigenvalue of the matrix $A$. Then

1. if $\operatorname{Re}\lambda > 0$, then all the corresponding solutions exponentially increase as $t \to \infty$,
2. if $\operatorname{Re}\lambda < 0$, then all the corresponding solutions exponentially decrease as $t \to \infty$,
3. if $\operatorname{Re}\lambda = 0$, then all the solutions are bounded in the case when only simple elementary divisors correspond to $\lambda$, and there are solutions growing as powers of $t$ in the opposite case.

REMARK 1.3.2 (on the estimate of the matriciant of system (1.3.1)). From (1.3.3) we have

$$
(1.3.6)\qquad \|e^{At}\| \leqslant \|S\| \cdot \|S^{-1}\| \cdot \|e^{Bt}\| \leqslant \|S\| \cdot \|S^{-1}\| \max_{1 \leqslant j \leqslant k} \|e^{J_{p_j}(\lambda_j)t}\|.
$$

Let $\alpha = \max_i \operatorname{Re}\lambda_i$, $i = 1, \dots, k$. The form of (1.3.4) implies that

$$
(1.3.7)\qquad \|e^{J_\nu(\lambda)t}\| \leqslant e^{\alpha t} \|e^{J_\nu(0)t}\|.
$$

The matrix $\exp J_\nu(0)t$ contains powers of $t$ and for $t \geqslant 0$ can be estimated in the following two ways:

a) for any $\varepsilon > 0$ there exists a constant $C_\varepsilon$ such that

$$
(1.3.8)\qquad \|e^{J_\nu(0)t}\| \leqslant C_\varepsilon e^{\varepsilon t}.
$$

Indeed,

$$
\|e^{J_\nu(0)t}\| = \|e^{J_\nu(0)t}\| e^{-\varepsilon t} e^{\varepsilon t}.
$$

Thus, for the constant $C_\varepsilon$ in (1.3.8) we have

$$C_\varepsilon = \max_{\mathbb{R}_+} \|e^{J_\nu(0)t}\| e^{-\varepsilon t}.$$

b)

$$\|e^{J_\nu(0)t}\| \leqslant D(1+t)^{\nu-1}. \tag{1.3.9}$$

Indeed, the power of $t$ on the left-hand side of the last inequality does not exceed $\nu - 1$, the constant $D \geqslant 1$ depends on the choice of the norm, and on the right-hand side of (1.3.9) there is the expression $(1+t)$, which guarantees the validity of the inequality for $t = 0$.

Therefore, the estimates a) and b), with (1.3.7) and (1.3.6) taken into account, generate the following estimates:

$$1.\quad \|e^{At}\| \leqslant C_\varepsilon e^{(\alpha+\varepsilon)t}, \qquad t \geqslant 0, \tag{1.3.10}$$

$$2.\quad \|e^{At}\| \leqslant K(1+t)^{n-1} e^{\alpha t}, \qquad t \geqslant 0. \tag{1.3.11}$$

In the last estimate the number $n$ may be replaced with the maximal order of Jordan blocks.

Note that in the case of simple eigenvalues of the matrix $A$ the estimate acquires the following form:

$$\|e^{At}\| \leqslant M e^{\alpha t}. \tag{1.3.12}$$

Let us consider some examples of the definition of $e^{At}$ and the estimate of $\|e^{At}\|$.

EXAMPLE 1.3.1.

$$\begin{aligned} \dot{x}_1 &= x_1 + 3x_2, \\ \dot{x}_2 &= 3x_1 - 2x_2 - x_3, \\ \dot{x}_3 &= -x_2 + x_3; \end{aligned} \qquad \text{here} \qquad A = \begin{pmatrix} 1 & 3 & 0 \\ 3 & -2 & -1 \\ 0 & -1 & 1 \end{pmatrix}.$$

The eigenvalues of the matrix $A$ are determined from the equation $\operatorname{Det}(A - \lambda E) = 0$ and are $\lambda_1 = 1$, $\lambda_2 = 3$, $\lambda_3 = -4$. The matrix $S$ consists of the eigenvectors of the matrix $A$. Hence,

$$S = \begin{pmatrix} 1 & 3 & 3 \\ 0 & 2 & -5 \\ 3 & -1 & -1 \end{pmatrix}, \qquad S^{-1} = \frac{1}{70}\begin{pmatrix} 7 & 0 & 21 \\ 15 & 10 & -5 \\ 6 & -10 & -2 \end{pmatrix}.$$

According to formula (1.3.2), $B = S^{-1}AS = \operatorname{diag}[1, 3, -4]$. Therefore, in accordance with (1.3.3), we have

$$X(t,0) = e^{At} = \frac{1}{70}\begin{pmatrix} 1 & 3 & 3 \\ 0 & 2 & -5 \\ 3 & -1 & -1 \end{pmatrix}\begin{pmatrix} e^t & 0 & 0 \\ 0 & e^{3t} & 0 \\ 0 & 0 & e^{-4t} \end{pmatrix}\begin{pmatrix} 7 & 0 & 21 \\ 15 & 10 & -5 \\ 6 & -10 & -2 \end{pmatrix}.$$

Note that, by virtue of expression (1.0.4), the matrix

$$X(t) = Se^{Bt} = \begin{pmatrix} 1 & 3 & 3 \\ 0 & 2 & -5 \\ 3 & -1 & -1 \end{pmatrix}\begin{pmatrix} e^t & 0 & 0 \\ 0 & e^{3t} & 0 \\ 0 & 0 & e^{-4t} \end{pmatrix}$$

is also fundamental, and, although the normalization condition at the point $t = 0$ is lost, this matrix is easier to calculate (we need not find $S^{-1}$).

Finally, we have

$$\|e^{At}\|_I \leqslant \|S\|_I \|S^{-1}\|_I \|e^{Bt}\|_I = \frac{7 \cdot 30}{70} e^{3t} = 3e^{3t}.$$

EXAMPLE 1.3.2.

$$\begin{aligned} \dot{x}_1 &= x_1 + x_2, \\ \dot{x}_2 &= -x_1 + 2x_2 + x_3, \\ \dot{x}_3 &= x_1 + x_3; \end{aligned} \qquad \text{here} \qquad A = \begin{pmatrix} 1 & 1 & 0 \\ -1 & 2 & 1 \\ 1 & 0 & 1 \end{pmatrix}.$$

The eigenvalues of the matrix $A$ are $\lambda_1 = 2$, $\lambda_{2,3} = 1 \pm i$. The matrix $S$ consists of eigenvectors and has the following form:

$$S = \begin{pmatrix} 1 & i & -i \\ 1 & -1 & -1 \\ 1 & 1 & 1 \end{pmatrix};$$

therefore,

$$S^{-1} = \frac{1}{4} \begin{pmatrix} 0 & 2 & 2 \\ -2i & -1+i & 1+i \\ 2i & -(1+i) & 1-i \end{pmatrix},$$

$$B = S^{-1}AS = \begin{pmatrix} 2 & 0 & 0 \\ 0 & 1+i & 0 \\ 0 & 0 & 1-i \end{pmatrix}.$$

This matrix has two complex conjugate Jordan blocks of order one, i.e., $\nu = 1$, and

$$S_{2\nu} = S_2 = \begin{pmatrix} 1 & i \\ 1 & -i \end{pmatrix} \qquad \text{and} \qquad S_2^{-1} = \frac{1}{2} \begin{pmatrix} 1 & 1 \\ -i & i \end{pmatrix}.$$

Thus, we have

$$\widetilde{B} = \begin{pmatrix} 1 & 0 & 0 \\ 0 & 1/2 & 1/2 \\ 0 & -i/2 & i/2 \end{pmatrix} \begin{pmatrix} 2 & 0 & 0 \\ 0 & 1+i & 0 \\ 0 & 0 & 1-i \end{pmatrix} \begin{pmatrix} 1 & 0 & 0 \\ 0 & 1 & i \\ 0 & 1 & -i \end{pmatrix} = \begin{pmatrix} 2 & 0 & 0 \\ 0 & 1 & -1 \\ 0 & 1 & 1 \end{pmatrix}.$$

By virtue of our notation, $\widetilde{B} = S_0^{-1} A S_0$. Then

$$S_0 = S \begin{pmatrix} 1 & 0 \quad 0 \\ 0 & \\ 0 & S_2 \end{pmatrix} = \begin{pmatrix} 1 & 0 & -2 \\ 1 & -2 & 0 \\ 1 & 2 & 0 \end{pmatrix},$$

$$S_0^{-1} = \begin{pmatrix} 1 & 0 \quad 0 \\ 0 & \\ 0 & S_2^{-1} \end{pmatrix} S^{-1} = \frac{1}{4} \begin{pmatrix} 0 & 2 & 2 \\ 0 & -1 & 1 \\ -2 & 1 & 1 \end{pmatrix}.$$

Finally, we have

$$e^{At} = S_0 e^{\widetilde{B}t} S_0^{-1} = S_0 \begin{pmatrix} e^{2t} & 0 & 0 \\ 0 & e^t \cos t & -e^t \sin t \\ 0 & e^t \sin t & e^t \cos t \end{pmatrix} S_0^{-1},$$

$$\|e^{At}\|_I \leqslant \|S\|_I \|S^{-1}\|_I \|e^{Bt}\|_I = \frac{3}{4}(2 + 2\sqrt{2})e^{2t} = \frac{3}{2}(1 + \sqrt{2})e^{2t}.$$

EXAMPLE 1.3.3.

$$\begin{aligned}\dot{x}_1 &= 8x_1 - x_2 - 5x_3,\\ \dot{x}_2 &= -2x_1 + 3x_2 + x_3,\\ \dot{x}_3 &= 4x_1 - x_2 - x_3;\end{aligned} \qquad \text{here} \qquad A = \begin{pmatrix} 8 & -1 & -5 \\ -2 & 3 & 1 \\ 4 & -1 & -1 \end{pmatrix}.$$

The eigenvalues of the matrix $A$ are $\lambda_1 = 2$, $\lambda_{2,3} = 4$.

$$S = \begin{pmatrix} 1 & 1 & 1/3 \\ 1 & -1 & 1/3 \\ 1 & 1 & 0 \end{pmatrix}, \qquad S^{-1} = \begin{pmatrix} -1/2 & 1/2 & 1 \\ 1/2 & -1/2 & 0 \\ 3 & 0 & -3 \end{pmatrix},$$

$$B = S^{-1}AS = \begin{pmatrix} 2 & 0 & 0 \\ 0 & 4 & 1 \\ 0 & 0 & 4 \end{pmatrix},$$

$$e^{At} = Se^{Bt}S^{-1} = S \begin{pmatrix} e^{2t} & 0 & 0 \\ 0 & e^{4t} & te^{4t} \\ 0 & 0 & e^{4t} \end{pmatrix} S^{-1}.$$

Let us estimate $\|e^{At}\|$.

1. The estimate according to formula (1.3.10):

$$\|e^{At}\|_I \leqslant \|S\|_I \|S^{-1}\|_I \|e^{Bt}\|_I = \frac{7}{3} 6e^{4t}(1+t) = 14e^{4t}(1+t)e^{-\varepsilon t}e^{\varepsilon t},$$

$$\max_{\mathbb{R}_+}(1+t)e^{-\varepsilon t} = \frac{1}{\varepsilon}e^{-\varepsilon(1-\varepsilon)/\varepsilon} = \frac{1}{\varepsilon}e^{\varepsilon-1};$$

therefore,

$$\|e^{At}\|_I \leqslant \frac{14}{\varepsilon}e^{\varepsilon-1}\varepsilon^{(4+\varepsilon)t} = M_\varepsilon e^{(4+\varepsilon)t}.$$

2. The estimate according to formula (1.3.11):

$$\|e^{At}\|_I \leqslant 14(1+t)e^{4t}.$$

## §4. Homogeneous systems with periodic coefficients. The Floquet theorem

We shall consider systems of the form

$$\dot{x} = A(t)x, \tag{1.4.1}$$

where $x \in \mathbb{C}^n$, $A(t+\omega) = A(t)$, $\omega > 0$, $A \in C(\mathbb{R})$.

Note some properties of the fundamental matrices of these systems.

1. If $X(t)$ is a fundamental matrix, then $X(t+\omega)$ is also a fundamental matrix. Indeed, by the condition we have

$$\frac{dX(t)}{dt} \equiv A(t)X(t), \qquad t \in \mathbb{R}.$$

Setting $t = t + \omega$ in this identity, we obtain

$$\frac{dX(t+\omega)}{dt} \equiv A(t+\omega)X(t+\omega) = A(t)X(t+\omega).$$

2. When the argument is augmented by the period, the fundamental matrix acquires a nonsingular matrix factor on the right.

Since the matrices $X(t)$ and $X(t+\omega)$ are fundamental, there exists a nonsingular matrix $B$ such that

$$X(t+\omega) = X(t)B.$$

3. The matrices $B$ corresponding to various fundamental matrices are similar.

Let $X(t)$ and $X_1(t)$ be two fundamental matrices. Then there exists a nonsingular matrix $C$ such that $X_1(t) = X(t)C$. Thus,

$$X_1(t+\omega) = X(t+\omega)C = X(t)BC = X(t)CC^{-1}BC = X_1(t)B_1.$$

4. The monodromy matrix. From the previous item it is clear that the eigenvalues and elementary divisors of the matrices $B$ are invariants of system (1.4.1). They define the dynamics of the solutions with the growth of $t$, since $X(t+m\omega) = X(t)B^m$. Let us study them in detail, using a specific matrix $B$ generated by the fundamental matrix normalized at $t = 0$. By property 2, $\Omega_0^{t+\omega} A = \Omega_0^t A \cdot B$, which implies $B = \Omega_0^{\omega} A$ for $t = 0$. This matrix is called the *monodromy matrix*.

5. Multipliers.

DEFINITION 1.4.1. The eigenvalues of the monodromy matrix are called *multipliers* of system (1.4.1).

The equation

$$\operatorname{Det}(B - \rho E) = 0,$$

defining the multipliers, is called *characteristic*. It follows from the last equality and the Ostrogradskiĭ-Liouville formula (1.0.5) that the product of the multipliers is $\operatorname{Det} B = \exp \int_0^{\omega} \operatorname{Sp} A(u)\, du$, i.e., no multiplier is zero.

The multipliers do not change if the system is subjected to a nonsingular $\omega$-periodic transformation $y = S(t)x$. Indeed, for fundamental matrices $X(t)$ and $Y(t)$ we have

$$Y(t+\omega) = S(t+\omega)X(t+\omega) = S(t)X(t)B = Y(t)B.$$

The origin of the term *multiplier*, i.e, a factor, is clarified by the following statement.

THEOREM 1.4.1. *A number $\rho$ is a multiplier of system* (1.4.1) *if and only if there esists a nontrivial solution $x(t)$ of this system such that*

$$x(t+\omega) = \rho x(t). \tag{1.4.2}$$

DEFINITION 1.4.2. A solution $x(t)$ satisfying identity (1.4.2) is called *normal*.

PROOF. *Necessity.* Let $\rho$ be a multiplier of system (1.4.1), i.e., let $\rho$ be an eigenvalue of the monodromy matrix $X(\omega)$, where $X(t) \equiv \Omega_0^t A$. The eigenvector $v$ corresponding to this $\rho$ satisfies the condition $X(\omega)v = \rho v$. Let us show that the solution $x(t) = X(t)v$ is the required one. Indeed,

$$x(t+\omega) = X(t+\omega)v = X(t)X(\omega)v = X(t)\rho v = \rho x(t).$$

*Sufficiency.* Let us show that the number from the identity (1.4.2) is a multiplier. For the soltuion $x(t)$ satisfying (1.4.2) we have

$$x(\omega) = \rho x(0) \qquad \text{for} \qquad t = 0.$$

Let us rewrite this solution as $x(t) = X(t)x(0)$; this implies

$$x(\omega) = X(\omega)x(0) \qquad \text{for} \qquad t = \omega.$$

From these two equalities it follows that

$$X(\omega)x(0) = \rho x(0).$$

Since $\|x(0)\| \neq 0$, we obtain $\mathrm{Det}[X(\omega) - \rho E] = 0$, i.e., $\rho$ is a multiplier. □

COROLLARY 1.4.1. *System* (1.4.1) *has an $\omega$-periodic solution if and only if one of its multipliers is equal to unity. To the multiplier $\rho = -1$ there corresponds an antiperiodic solution $x(t+\omega) = -x(t)$ of period $2\omega$.*

REMARK 1.4.1 (on the structure of a normal solution). Let us set $\rho = \exp \lambda\omega$ and rewrite a normal solution in the form $x(t) = (\exp \lambda t)\varphi(t)$. We show that $\varphi(t+\omega) = \varphi(t)$. Indeed,

$$\begin{aligned}\varphi(t+\omega) &= e^{-\lambda(t+\omega)}x(t+\omega) = e^{-\lambda(t+\omega)}\rho x(t)\\ &= e^{-\lambda(t+\omega)}e^{\lambda\omega}e^{\lambda t}\varphi(t) = \varphi(t).\end{aligned}$$

Before passing to the study of the structure of the fundamental matrix of system (1.4.1), we introduce the notion of the logarithm of a matrix.

DEFINITION 1.4.3. If the equality

$$e^Y = X$$

is valid for two square matrices $X$ and $Y$, then the matrix $Y$ is called the *logarithm of the matrix* $X$ and is denoted by $Y = \mathrm{Ln}\, X$.

LEMMA 1.4.1. *Any nonsingular matrix $X$ has a logarithm.*

PROOF. First, let $X = J_l(\lambda)$ be a Jordan block, $\lambda \neq 0$. We write it in the form

$$X = \lambda E_l + J_l(0) = \lambda\left[E_l + \frac{J_l(0)}{\lambda}\right].$$

For the scalar logarithmic function we have

$$\ln(1+z) = \sum_{k=1}^{\infty}(-1)^{k-1}\frac{z^k}{k}, \qquad |z| < 1.$$

By analogy, let us consider the matrix series

$$E_l \,\mathrm{Ln}\,\lambda + \sum_{k=1}^{\infty}(-1)^{k-1}\frac{1}{k\lambda^k}J_l^k(0), \tag{1.4.3}$$

where

$$\mathrm{Ln}\,\lambda = \ln|\lambda| + i(\arg\lambda + 2k\pi), \qquad k = 0, \pm 1, \ldots.$$

We show that the series (1.4.3) converges and that its sum satisfies the definition of the matrix logarithm.

$J_l(0)$ is a nilpotent matrix; therefore, (1.4.3) is a finite sum. Let us denote it by $Y$. Thus, $\exp Y$ is well defined. Exponential series have identical expansions for scalar and matrix quantities; hence,

$$\begin{aligned} e^Y &= \exp\left(E_l \operatorname{Ln}\lambda + \sum_{k=1}^{\infty}(-1)^{k-1}\frac{1}{k\lambda^k}J_l^k(0)\right) \\ &= e^{\operatorname{Ln}\lambda}\left(E_l + \frac{J_l(0)}{\lambda}\right) = \lambda E_l + J_l(0) = X. \end{aligned}$$

Thus,

$$Y = \operatorname{Ln} X = \operatorname{Ln} J_l(\lambda) = E_l \operatorname{Ln}\lambda + \sum_{k=1}^{l-1}\frac{(-k)^{k-1}}{k\lambda^k}J_l^k(0).$$

Now let us pass to the case of an arbitrary nonsingular matrix $X$. It can always be written in the form

$$X = S\operatorname{diag}[J_{p_1}(\lambda_1), J_{p_2}(\lambda_2), \dots, J_{p_k}(\lambda_k)]S^{-1}$$

(see formula (1.3.2)).

Let us consider the matrix

$$Y = S\operatorname{diag}[\operatorname{Ln} J_{p_1}(\lambda_1), \dots, \operatorname{Ln} J_{p_k}(\lambda_k)]S^{-1} \tag{1.4.4}$$

and verify that $Y = \operatorname{Ln} X$.

Indeed,

$$e^Y = S\operatorname{diag}[e^{\operatorname{Ln} J_{p_1}(\lambda_1)}, \dots, e^{\operatorname{Ln} J_{p_k}(\lambda_k)}]S^{-1} = X. \quad \square$$

Let us discuss the theorem proved.

REMARK 1.4.2. The fact that $\operatorname{Ln}\lambda$ is multivalued implies that $\operatorname{Ln} J_l(\lambda)$ is multivalued; therefore, $\operatorname{Ln} X$ is multivalued.

REMARK 1.4.3. Let $\lambda_1, \lambda_2, \dots, \lambda_n$ be the eigenvalues of the matrix $X$. As we can see from (1.4.4), under a specific choice of the branch, $\operatorname{Ln}\lambda_1, \operatorname{Ln}\lambda_2, \dots, \operatorname{Ln}\lambda_n$ are eigenvalues of the matrix $\operatorname{Ln} X$. The structure of the Jordan canonical form of the matrices $X$ and $\operatorname{Ln} X$ is identical. The latter statement follows from the fact that the triangular matrix $\operatorname{Ln} J_l(\lambda)$ is similar to a Jordan block of the same size **[16]**.

REMARK 1.4.4. We note the cases when there exists a real logarithm for a real matrix $X$.

1. All the eigenvalues of the matrix $X$ are positive. In this case their logarithms may be chosen real, in which case $\operatorname{Ln} X$ is real (see (1.4.4)).

2. Among the eigenvalues of the matrix $X$ some are complex but there are no negative ones. We choose the logarithms of the complex conjugate eigenvalues so that they are complex conjugate and then we have in (1.4.4) that the matrices $S$, $S^{-1}$ are complex and the blocks of the diagonal matrix are either real or pairwise complex conjugate. Further, we carry out two more transformations of the block-diagonal matrix, i.e., first, we reduce it to canonical form and then by means of the transformation $S_{2\nu}$ (see §3), we render real the pairwise complex conjugate Jordan blocks. As a result, expression (1.4.4) will take the form $Y = \widetilde{S}Y_1\widetilde{S}^{-1}$, where $Y_1$ is

already a real matrix and the matrix $\widetilde{S}$ is the product of the three matrices corresponding to three consecutive transformations applied. Consider

$$X_1 = e^{Y_1} = \widetilde{S}^{-1} e^Y \widetilde{S} = \widetilde{S}^{-1} X \widetilde{S}.$$

The matrices $X_1$ and $X$ are real and similar; thus, by Lemma 1.3.1, there exists a real matrix $V$ such that $X = VX_1V^{-1} = Ve^{Y_1}V^{-1} = e^{VY_1V}$; this implies that the matrix $VY_1V^{-1}$ is the real logarithm of the matrix $X$.

3. The negative eigenvalues of the matrix $X$ are such that an even number of identical Jordan blocks corresponds to each $\lambda < 0$. Then in (1.4.4) in one half of the blocks we take

$$\operatorname{Ln}\lambda = \ln|\lambda| + i\pi,$$

and in the other

$$\operatorname{Ln}\lambda = \ln|\lambda| + i(\pi - 2\pi),$$

i.e., we obtain pairwise complex conjugate blocks and reduce the situation to that of item 2.

In **[16]** the following result is proved.

THEOREM 1.4.2. *A real nonsingular matrix $X$ has a real logarithm if and only if the matrix either has no elementary divisors corresponding to negative eigenvalues, or each such divisor is of even multiplicity.*

COROLLARY 1.4.2. *The matrix $\operatorname{Ln} X^2$ can always be chosen so that it is real for any real matrix $X$.*

Let us consider examples of the definition of the logarithm of a matrix.

EXAMPLE 1.4.1.

$$A = \begin{pmatrix} 1 & 3 & 0 \\ 3 & -2 & -1 \\ 0 & -1 & 1 \end{pmatrix};$$

the eigenvalues of $A$ are

$$\lambda_1 = 1, \qquad \lambda_2 = 3, \qquad \lambda_3 = -4.$$

By equality (1.4.4), we have one of the logarithms:

$$\operatorname{Ln} A = S \operatorname{diag}[0, \ln 3, \ln 4 + i\pi] S^{-1}.$$

The matrices $S$ and $S^{-1}$ were found in Example 1.3.1. There is no real logarithm.

EXAMPLE 1.4.2.

$$A = \begin{pmatrix} 8 & -1 & -5 \\ -2 & 3 & 1 \\ 4 & -1 & -1 \end{pmatrix};$$

the eigenvalues of $A$ are

$$\lambda_1 = 2, \qquad \lambda_{2,3} = 4.$$

The canonical forms of the matrices $B$, $S$, and $S^{-1}$ were found in Example 1.3.3. The notation (1.4.4) implies that we can choose the real logarithm

$$\operatorname{Ln} A = S \begin{pmatrix} \ln 2 & 0 & 0 \\ 0 & \ln 4 & 1/4 \\ 0 & 0 & \ln 4 \end{pmatrix} S^{-1}.$$

EXAMPLE 1.4.3.

$$A = \begin{pmatrix} 1 & 1 & 0 \\ -1 & 2 & 1 \\ 1 & 0 & 1 \end{pmatrix};$$

the eigenvalues of $A$ are

$$\lambda_1 = 2, \qquad \lambda_{2,3} = 1 \pm i.$$

From the expression (1.4.4) we have, e.g.,

$$\operatorname{Ln} A = S \begin{pmatrix} \ln 2 & 0 & 0 \\ 0 & \ln\sqrt{2} + i\pi/4 & 0 \\ 0 & 0 & \ln\sqrt{2} - i\pi/4 \end{pmatrix} S^{-1}.$$

The matrices $S$ and $S^{-1}$ were found in Example 1.3.2. The logarithm obtained is complex; by Remark 1.4.4, item 2, the matrix $A$ has a real logarithm. Let us obtain it by means of a transformation of the type $S_{2\nu}$.

In our case (see Example 1.3.2)

$$\begin{aligned} \operatorname{Ln} A = {} & S \begin{pmatrix} 1 & 0 \; 0 \\ 0 & \\ 0 & S_2 \end{pmatrix} \begin{pmatrix} 1 & 0 \; 0 \\ 0 & \\ 0 & S_2^{-1} \end{pmatrix} \begin{pmatrix} \ln 2 & 0 & 0 \\ 0 & \ln\sqrt{2} + i\pi/4 & 0 \\ 0 & 0 & \ln\sqrt{2} - i\pi/4 \end{pmatrix} \\ & \times \begin{pmatrix} 1 & 0 \; 0 \\ 0 & \\ 0 & S_2 \end{pmatrix} \begin{pmatrix} 1 & 0 \; 0 \\ 0 & \\ 0 & S_2^{-1} \end{pmatrix} S^{-1} \\ = {} & S_0 \begin{pmatrix} \ln 2 & 0 & 0 \\ 0 & \ln\sqrt{2} & -\pi/4 \\ 0 & \pi/4 & \ln\sqrt{2} \end{pmatrix} S_0^{-1}. \end{aligned}$$

EXAMPLE 1.4.4.

$$A = \begin{pmatrix} -3 & 0 \\ 0 & -3 \end{pmatrix}.$$

Consider

$$\operatorname{Ln} A = \begin{pmatrix} \ln 3 + i\pi & 0 \\ 0 & \ln 3 - i\pi \end{pmatrix}.$$

Applying the transformation $S_2$, we obtain

$$\begin{aligned} X_1 &= S_2^{-1} \operatorname{Ln} A S_2 \\ &= \frac{1}{2} \begin{pmatrix} 1 & 1 \\ -i & i \end{pmatrix} \begin{pmatrix} \ln 3 + i\pi & 0 \\ 0 & \ln 3 - i\pi \end{pmatrix} \begin{pmatrix} 1 & i \\ 1 & -1 \end{pmatrix} = \begin{pmatrix} \ln 3 & -\pi \\ \pi & \ln 3 \end{pmatrix}. \end{aligned}$$

This implies that the matrix

$$A_1 = e^{X_1} = S_2^{-1} A S_2$$

is real and similar to the matrix $A$, which is also real. By Lemma 1.3.1, there exists a real nonsingular matrix $V$ such that

$$A = V A_1 V^{-1} = V e^{X_1} V^{-1} = e^{V X_1 V^{-1}}.$$

The matrix $V X_1 V^{-1}$ is a real logarithm of the matrix $A$.

Let us return to system (1.4.1). We have the following statement.

THEOREM 1.4.3 (Floquet). *The matriciant of system* (1.4.1) *can be represented in the form*

$$\underset{0}{\overset{t}{\Omega}} A = \Phi(t) e^{\Lambda t}, \tag{1.4.5}$$

*where* $\Lambda = \frac{1}{\omega} \operatorname{Ln} \Omega_0^\omega A$, $\Phi(t) = \Phi(t+\omega)$.

PROOF. We have the identity

$$\underset{0}{\overset{t}{\Omega}} A \equiv \underset{0}{\overset{t}{\Omega}} A e^{-\Lambda t} e^{\Lambda t} \equiv \Phi(t) e^{\Lambda t}.$$

Let us verify that the matrix $\Phi(t)$ is $\omega$-periodic:

$$\Phi(t+\omega) = \underset{0}{\overset{t+\omega}{\Omega}} \cdot A e^{-\Lambda(t+\omega)} = \underset{0}{\overset{t}{\Omega}} A \underset{0}{\overset{\omega}{\Omega}} A e^{-\Lambda\omega} e^{-\Lambda t} = \Phi(t). \quad \square$$

REMARK 1.4.5. Any fundamental matrix $X_1(t)$ of system (1.4.1) can be represented in the form (1.4.5) with corresponding matrices $\Phi_1(t)$ and $\Lambda_1$.

Indeed, there exists a nondegenerate matrix $C$ such that

$$X_1(t) = \underset{0}{\overset{t}{\Omega}} A C;$$

therefore,

$$X_1(t) = \Phi(t) e^{\Lambda t} C = \Phi(t) C C^{-1} e^{\Lambda t} C \equiv \Phi_1(t) e^{\Lambda_1 t}, \qquad \Lambda_1 = C^{-1} \Lambda C.$$

REMARK 1.4.6. The matrix

$$\Lambda = \frac{1}{\omega} \operatorname{Ln} \underset{0}{\overset{\omega}{\Omega}} A$$

and, therefore, $\Phi(t)$ are, generally speaking, complex. In the case when $A(t)$ is real, the matrix $\Lambda$ can be chosen real if and only if among the multipliers of the system there are either no negative ones, or every elementary divisor corresponding to negative multipliers is of even multiplicity.

REMARK 1.4.7. In the representation of the type (1.4.5) one can avoid complex quantities if one requires that the matrix $\Phi(t)$ be $2\omega$-periodic. Indeed,

$$\underset{0}{\overset{t+2\omega}{\Omega}} A = \underset{0}{\overset{t+\omega}{\Omega}} A \underset{0}{\overset{\omega}{\Omega}} A = \underset{0}{\overset{t}{\Omega}} A \left( \underset{0}{\overset{\omega}{\Omega}} A \right)^2.$$

By Corollary 1.4.2, there exists a real matrix

$$\widetilde{\Lambda} = \frac{1}{2\omega} \operatorname{Ln} \underset{0}{\overset{2\omega}{\Omega}} A.$$

Then the matrix

$$\widetilde{\Phi}(t) = \underset{0}{\overset{t}{\Omega}} A \cdot e^{-\widetilde{\Lambda} t}$$

is real for real $A(t)$, and $\widetilde{\Phi}(t+2\omega) = \widetilde{\Phi}(t)$; finally,

$$\underset{0}{\overset{t}{\Omega}} A = \widetilde{\Phi}(t) e^{\widetilde{\Lambda} t}.$$

REMARK 1.4.8 (on the behavior of the solutions of system (1.4.1) as $t \to \infty$). From the representation of the matriciant (1.4.5) it is clear that the dynamics of the solutions is determined by the matrix $\exp \Lambda t$, which is fundamental for the system $\dot{x} = \Lambda x$. The eigenvalues $\lambda_1, \lambda_2, \dots, \lambda_n$ of the matrix $\Lambda$ are connected with the multipliers $\rho_1$, $\rho_2, \dots, \rho_n$ of system (1.4.1) in the following way:

$$\lambda_k = \frac{1}{\omega} \operatorname{Ln} \rho_k = \frac{1}{\omega}[\ln |\rho_k| + i(\arg \rho_k + 2m\pi)], \qquad m = 0, \pm 1, \dots, \qquad k = 1, 2, \dots, n,$$

and the elementary divisors corresponding to $\lambda_k$ and $\rho_k$ coincide. This and the statements of Remark 1.3.1 imply that for a multiplier $\rho$ of system (1.4.1) the following is valid:

1) if $|\rho| > 1$, then all the corresponding solutions exponentially increase as $t \to \infty$,

2) if $|\rho| < 1$, then all the corresponding solutions exponentially decrease as $t \to \infty$,

3) if $|\rho| = 1$, then, in the case when only simple elementary divisors correspond to $\rho$, all the solutions are bounded, and there exist solutions growing as powers of $t$ if among the elementary divisors there are multiple ones.

REMARK 1.4.9 (on the multipliers of the adjoint system). Let $X(\omega)$ be the monodromy matrix of system (1.4.1) and let $\rho_1, \rho_2, \dots, \rho_n$ be its multipliers. For the adjoint system $\dot{x} = -A^*(t)x$ the monodromy matrix is the matrix $[X^{-1}(\omega)]^*$ (see Theorem 1.1.1) and, therefore, the multipliers are the quantities $\bar{\rho}_1^{-1}, \bar{\rho}_2^{-1}, \dots, \bar{\rho}_n^{-1}$, i.e., the multipliers of the initial and adjoint systems are symmetric with respect to the unit circle. Hence, e.g., a group of exponentially decreasing solutions of the adjoint system corresponds to a group of exponentially increasing solutions of system (1.4.1), and vice versa.

## §5. Nonhomogeneous periodic systems

Naturally, for periodic in $t$ linear nonhomogeneous systems

$$\dot{y} = A(t)y + f(t), \tag{1.5.1}$$

where

$$y \in \mathbb{C}^n, \qquad A, f \in C(\mathbb{R}),$$
$$f(t+\omega) = f(t), \qquad A(t) = A(t+\omega), \qquad \omega > 0, \qquad t \in \mathbb{R},$$

there arise questions whether $\omega$-periodic solutions exist and if they do, then what is their number, and what properties of systems (1.4.1) and (1.5.1) determine this. The following three theorems provide a partial answer to these questions.

First let us prove the following result.

LEMMA 1.5.1. *A solution $y(t)$ of system (1.5.1) is $\omega$-periodic if and only if $y(0) = y(\omega)$.*

PROOF. Necessity is obvious. Let us prove sufficiency. Thus, $y(t)$ is a solution of system (1.5.1) such that $y(0) = y(\omega)$. Note that the vector-function $\varphi(t) = y(t+\omega)$ is also a solution of system (1.5.1) because the right-hand side is $\omega$-periodic in $t$. Both solutions take the same value at $t = 0$ since $y(0) = y(\omega) = \varphi(0)$; thus, they coincide identically, i.e., $y(t) \equiv y(t+\omega)$ for $t \in \mathbb{R}$. □

THEOREM 1.5.1. *If a linear homogeneous system (1.4.1) has no nontrivial $\omega$-periodic solutions, then the corresponding nonhomogeneous system (1.5.1) has a unique $\omega$-periodic solution.*

PROOF. Let $X(t)$ be a fundamental matrix of system (1.4.1) and $X(0) = E$. We show that under our assumption in the set of solutions

$$y(t) = X(t)y(0) + \int_0^t X(t,\tau)f(\tau)\,d\tau \tag{1.5.2}$$

of (1.5.1) there is a unique $\omega$-periodic $\widetilde{y}(t)$. By Lemma 1.5.1, a solution $\widetilde{y}(t)$ is periodic if and only if $\widetilde{y}(0)$ satisfies the condition

$$\widetilde{y}(0) = X(\omega)\widetilde{y}(0) + \int_0^\omega X(\omega,\tau)f(\tau)\,d\tau,$$

or

$$[E - X(\omega)]\widetilde{y}(0) = \int_0^\omega X(\omega,\tau)f(\tau)\,d\tau. \tag{1.5.3}$$

By the condition of the theorem, there are no multipliers (i.e., eigenvalues of the matrix $X(\omega)$) equal to unity. Hence, $\operatorname{Det}[X(\omega) - E] \neq 0$ and equality (1.5.3) determines a unique vector $\widetilde{y}(0)$. The corresponding $\omega$-periodic solution from the set (1.5.2) has the form

$$y(t) = X(t)[E - X(\omega)]^{-1}\int_0^\omega X(\omega,\tau)f(\tau)\,d\tau + \int_0^t X(t,\tau)f(\tau)\,d\tau. \quad \square \tag{1.5.4}$$

REMARK 1.5.1. If a linear homogeneous system has an $\omega$-periodic solution, then the corresponding linear nonhomogeneous system may not have a solution of this type.

A simple illustration of this statement is the case of resonance.

The general solution of the equation $y'' + y = \sin t$ has the form

$$y = A\sin(t+\delta) - \frac{t}{2}\cos t$$

and this equation has no $2\pi$-periodic solutions, while any solution of the corresponding homogeneous equation is $2\pi$-periodic.

The following theorem provides conditions under which both the homogeneous and nonhomogeneous systems have $\omega$-periodic solutions.

THEOREM 1.5.2. *Let a linear homogeneous $\omega$-periodic system* (1.4.1) *have $k \leqslant n$ linearly independent $\omega$-periodic solutions $\varphi_1(t), \varphi_2(t), \dots, \varphi_k(t)$. Then*

1. *The adjoint system*

$$\dot{z} = -A^*(t)z \tag{1.5.5}$$

*also has $k$ linearly independent $\omega$-periodic solutions $\psi_1(t), \dots, \psi_k(t)$.*

2. *The corresponding nonhomogeneous system* (1.5.1) *has $\omega$-periodic solutions if and only if the orthogonality conditions*

$$\int_0^\omega (\psi_s(t), f(t))\,dt = 0, \qquad s = 1, \dots, k, \tag{1.5.6}$$

*are satisfied and in this case the $\omega$-periodic solutions form a $k$-parametric family.*

PROOF. 1. Let $X(t)$ be the fundamental matrix of system (1.4.1) such that $X(0) = E$, and let $\varphi(t)$ be an $\omega$-periodic solution. According to Lemma 1.5.1, from the identity $\varphi(t) = X(t)\varphi(0)$ for $t = \omega$ we have

$$\varphi(0) = X(\omega)\varphi(0),$$

or

$$(X(\omega) - E)\varphi(0) = 0. \tag{1.5.7}$$

By the condition, this system has $k$ linearly independent solutions $\varphi_1(0), \dots, \varphi_k(0)$, i.e., $\operatorname{rank}(X(\omega) - E) = n - k$.

The monodromy matrix of the adjoint system is the matrix $Z(\omega) = [X^{-1}(\omega)]^*$ (see Theorem 1.1.1), and, by (1.5.7), a solution $\psi(t)$ is $\omega$-periodic if and only if its initial condition $\psi(0)$ satisfies the equation

$$([X^{-1}(\omega)]^* - E)\psi(0) = 0,$$

or

$$(X(\omega) - E)^*\psi(0) = 0. \tag{1.5.8}$$

Since $\operatorname{rank}(X(\omega) - E) = n - k$, (1.5.8) has $k$ linearly indepenent solutions

$$\psi_1(0), \dots, \psi_k(0),$$

which provide the initial data for $k$ $\omega$-periodic solutions of system (1.5.5).

2. Let us show that the conditions (1.5.6) are necessary for a periodic solution $y(t)$ of system (1.5.1) to exist. From equality (1.5.3) we have

$$[E - X(\omega)]y(0) = X(\omega)\int_0^\omega X^{-1}(\tau)f(\tau)\,d\tau, \tag{1.5.9}$$

and it follows from (1.5.8) that

$$[E - X(\omega)]^*\psi_s(0) = 0 \tag{1.5.10}$$

for any $\omega$-periodic solution $\psi_s(t)$, $s = 1, \dots, k$, of system (1.5.5). Using the two last equalities, we obtain

$$\begin{aligned}
0 &= ([E - X(\omega)]^*\psi_s(0), y(0))\\
&= (\psi_s(0), (E - X(\omega))y(0)) \overset{(1.5.9)}{=} \left(\psi_s(0), X(\omega)\int_0^\omega X^{-1}(\tau)f(\tau)\,d\tau\right)\\
&= \left(X^*(\omega)\psi_s(0), \int_0^\omega X^{-1}(\tau)f(\tau)\,d\tau\right) \overset{(1.5.10)}{=} \left(\psi_s(0), \int_0^\omega X^{-1}(\tau)f(\tau)\,d\tau\right)\\
&= \int_0^\omega (\psi_s(0), X^{-1}(\tau)f(\tau))\,d\tau = \int_0^\omega ([X^{-1}(\tau)]^*\psi_s(0), f(\tau))\,d\tau\\
&= \int_0^\omega (\psi_s(\tau), f(\tau))\,d\tau.
\end{aligned}$$

3. Let us prove that the conditions (1.5.6) are sufficient for the existence of an $\omega$-periodic solution of the nonhomogeneous system (1.5.1), which, in turn, is equivalent to the existence of a solution of system (1.5.3). To prove the latter fact let us show that the rank of the matrix of the coefficients of system (1.5.3) is equal to the rank of the augmented matrix.

Let $z_0$ satisfy the system

$$(X^*(\omega) - E)z_0 = 0. \tag{1.5.11}$$

It follows from (1.5.8) that $z_0$ is the initial condition of some $\omega$-periodic solution $z(t)$ of the adjoint system; therefore,

$$z(t) = [X^{-1}(t)]^* z_0 = \sum_{s=1}^{k} c_s \psi_s(t).$$

From this and (1.5.6) we have

$$\begin{aligned} 0 &= \int_0^\omega ([X^{-1}(t)]^* z_0, f(t))\, dt \\ &= \int_0^\omega (z_0, X^{-1}(t) f(t))\, dt \overset{(1.5.11)}{=} \int_0^\omega ([X(\omega)]^* z_0, X^{-1}(t) f(t))\, dt \\ &= \int_0^\omega (z_0, X(\omega) X^{-1}(t) f(t)) dt = (z_0, \int_0^\omega X(\omega) X^{-1}(t) f(t)\, dt). \end{aligned}$$

Thus, taking into account the conditions (1.5.6), we see that system (1.5.11) is equivalent to the system

$$\begin{aligned} (X(\omega) - E)^* z_0 = 0, \\ (z_0, \int_0^\omega X(\omega) X^{-1}(t) f(t)\, dt) = 0, \end{aligned}$$

which proves the statement on the equality of the ranks of matrices, i.e., system (1.5.3) is compatible and there exists an $\omega$-periodic solution $y(t)$ of the nonhomogeneous system. The complete set of its $\omega$-periodic solutions has the form

$$\sum_{s=1}^{k} c_s \psi_s(t) + y(t),$$

and is a $k$-parametric family. □

THEOREM 1.5.3 (Massera). *If a linear nonhomogeneous $\omega$-periodic system has a bounded solution $\tilde{y}(t)$ for $t \geqslant 0$, then it also has an $\omega$-periodic solution.*

PROOF. Let $X(t)$ be a fundamental matrix of system (1.4.1) and $X(0) = E$. The bounded solution $\tilde{y}(t)$ of system (1.5.1) has the following form:

$$\tilde{y}(t) = X(t)\tilde{y}(0) + \int_0^t X(t) X^{-1}(\tau) f(\tau)\, d\tau. \tag{1.5.12}$$

This implies that for $t = \omega$

$$\tilde{y}(\omega) = X(\omega)\tilde{y}(0) + b, \qquad \text{where} \qquad b = \int_0^\omega X(\omega, \tau) f(\tau)\, d\tau.$$

Note that if $b = 0$, then, according to (1.5.12), we have the required $\omega$-periodic solution with the initial condition $\tilde{y}(0) = 0$, i.e.,

$$\tilde{y}(t) = \int_0^t X(t, \tau) f(\tau)\, d\tau.$$

We proceed with the arguments assuming that $b \neq 0$.

The vector-function $\widetilde{y}(t+\omega)$ is also a solution of system (1.5.1) and

$$\widetilde{y}(t+\omega) = X(t)\widetilde{y}(\omega) + \int_0^t X(t)X^{-1}(\tau)f(\tau)\,d\tau.$$

Therefore,

$$\widetilde{y}(2\omega) = X(\omega)[X(\omega)\widetilde{y}(0) + b] + b = X^2(\omega)\widetilde{y}(0) + X(\omega)b + b.$$

By induction, we obtain

(1.5.13) $$\widetilde{y}(m\omega) = X^m(\omega)\widetilde{y}(0) + \sum_{k=0}^{m-1} X^k(\omega)b, \qquad m \in \mathbb{Z}_+.$$

Now let us assume that system (1.5.1) has no $\omega$-periodic solutions; thus, the system

(1.5.14) $$[E - X(\omega)]y_0 = b$$

is incompatible; therefore,

$$\operatorname{rank}[E - X(\omega)] = r < \operatorname{rank}[E - X(\omega), b].$$

This implies that the vector $b$ does not belong to the linear hull of $r$ linearly independent column vectors of $E - X(\omega)$ and, thus, has a component

$$\beta, \qquad (b, \beta) \neq 0,$$

in the $(n-r)$-dimensional orthogonal complement whose basis is determined by the system

$$[E - X(\omega)]^* c = 0.$$

Thus, we have

$$[E - X(\omega)]^* \beta = 0$$

and

(1.5.15) $$\beta = [X^k(\omega)]^* \beta.$$

Now we multiply (1.5.13) scalarly by $\beta$ and, taking into account (1.5.15), we write

$$\begin{aligned}(\widetilde{y}(m\omega), \beta) &= (\widetilde{y}(0), [X^m(\omega)]^*\beta) + \sum_{k=0}^{m-1}(b, [X^k(\omega)]^*\beta) \\ &= (\widetilde{y}(0), \beta) + m(b, \beta).\end{aligned}$$

Hence,

$$(\widetilde{y}(m\omega), \beta) \to \infty, \qquad m \to \infty,$$

which contradicts the boundedness of $\widetilde{y}(t)$. Therefore, system (1.5.14) is compatible and defines the initial condition of an $\omega$-periodic solution of system (1.5.1). □

CHAPTER II

# Lyapunov Characteristic Exponents in the Theory of Linear Systems

One of the main problems of the theory of differential equations is the behavior of solutions when the growth of the independent variable is unbounded. In the case of linear homogeneous systems with constant coefficients the answer depends on the signs of the real parts of the eigenvalues of the matrix of coefficients and the multiplicities of its elementary divisors (Remark 1.3.1), and in the case of periodic coefficients, on the location of the multipliers on the complex plane with respect to the unit circle and the multiplicities of the corresponding elementary divisors (Remark 1.4.8). In other cases the information available is not so detailed; however, there exist rather general results obtained by the method of characteristic exponents due to Lyapunov **[25]**. In the framework of this method, the growth rate of solutions is studied in comparison with the exponential functions $\exp \alpha t$. This growth is determined by characteristic exponents. The real parts of the eigenvalues of the matrices of coefficients are the characteristic exponents for the solutions of linear homogeneous systems with constant coefficients, and the real parts of the logarithms of the multipliers divided by the period, in the case of periodic coefficients.

## §1. Definition and main properties of characteristic exponents

Let a complex-valued function $f(t)$ be defined on the interval $[t_0, \infty)$.

DEFINITION 2.1.1. The number (or the symbol $\pm\infty$) defined as

$$\chi[f] = \overline{\lim_{t\to\infty}} \frac{1}{t} \ln |f(t)| \tag{2.1.1}$$

is called the *characteristic exponent* of the function $f(t)$.

The characteristic exponent determines the growth of the absolute value of a function with respect to the scale of exponential functions $\exp \alpha t$. Ovbiously, the characteristic exponent for the latter is the number $\alpha$. For an arbitrary function $f(t)$ the identity

$$|f(t)| = \exp\left(\frac{1}{t} \ln |f(t)|\right) t$$

is valid; this clarifies the definition given above.

We note that in Lyapunov's definition **[25]** the characteristic exponent differs by a sign from ours. The definition given above is standard in contemporary mathematics.

As an excercise, let us find the characteristic exponents of specific functions borrowed from **[25]**:

$$\chi[t^m] = 0, \qquad \chi[\exp(\pm t \sin t)] = 1,$$
$$\chi[c \neq 0] = 0, \qquad \chi[\exp(-t \exp \sin t)] = -e^{-1},$$
$$\chi[0] = -\infty, \qquad \chi[\exp(t \exp \sin t)] = e,$$
$$\chi\left[\exp\left(t \cos \frac{1}{t}\right)\right] = 1, \qquad \chi[t^t] = \infty,$$
$$\chi\left[\exp\left(-t \cos \frac{1}{t}\right)\right] = -1, \qquad \chi[t^{-1}] = -\infty.$$

Further we shall deal mainly with finite characteristic exponents with the exception of $\chi[0] = -\infty$.

Note the obvious properties following straightforwardly from Definition 2.1.1:

1. $\chi[f] = \chi[|f|]$,
2. $\chi[cf] = \chi[f], \quad |c| \neq 0$,
3. if $|f(t)| \leqslant |F(t)|$ for $t \geqslant a$, then

$$\chi[f] \leqslant \chi[F].$$

The next lemma provides more precise information on the growth of a function having a finite characteristic exponent.

LEMMA 2.1.1. *$\chi[f] = \alpha \neq \pm\infty$ if and only if for any $\varepsilon > 0$ the following two conditions hold simultaneously*:

$$\text{(2.1.2)} \qquad 1. \quad \lim_{t\to\infty} \frac{|f(t)|}{\exp(\alpha+\varepsilon)t} = 0,$$

$$\text{(2.1.3)} \qquad 2. \quad \overline{\lim_{t\to\infty}} \frac{|f(t)|}{\exp(\alpha-\varepsilon)t} = \infty.$$

PROOF. *Necessity.* Let

$$\text{(2.1.4)} \qquad \chi[f] = \overline{\lim_{t\to\infty}} \frac{1}{t} \ln |f(t)| = \alpha.$$

It follows from (2.1.4) that for a fixed $\varepsilon > 0$ there exists a $T > 0$ such that for $t > T$ the inequality

$$\frac{1}{t} \ln |f(t)| < \alpha + \frac{\varepsilon}{2}$$

is valid. In other words,

$$|f(t)| < \exp\left(\alpha + \frac{\varepsilon}{2}\right) t.$$

Therefore, we have

$$\lim_{t\to\infty} \frac{|f(t)|}{\exp(\alpha+\varepsilon/2)t\,\exp((\varepsilon/2)t)} \leqslant \lim_{t\to\infty} \exp(-\varepsilon/2)t = 0,$$

i.e., the condition (2.1.2) is valid.

Let a sequence $t_k \to \infty$, $k \to \infty$, realize equality (2.1.4); then there exists $N > 0$ such that for $k > N$ we have

$$\ln |f(t_k)| > (\alpha - \varepsilon/2) t_k$$

or

$$|f(t_k)| > \exp(\alpha - \varepsilon/2)t_k.$$

Thus, we have

$$\lim_{k\to\infty} \frac{|f(t_k)|}{\exp(\alpha-\varepsilon)t_k} = \lim_{k\to\infty}\left(\frac{|f(t_k)|}{\exp(\alpha-\varepsilon/2)t_k}\exp(\varepsilon/2)t_k\right)$$
$$\geqslant \lim_{k\to\infty}\exp(\varepsilon/2)t_k = \infty,$$

which proves (2.1.3).

*Sufficiency.* From (2.1.2) for sufficiently large $t$ there follows the inequality $|f(t)| < \exp(\alpha+\varepsilon)t$ and, since $\varepsilon > 0$ is arbitrary, we have

$$\chi[f] \leqslant \alpha.$$

Now let a sequence $t_k \to \infty$, $k \to \infty$, realize (2.1.3). Hence, for sufficiently large $k$ the inequality

$$|f(t_k)| > \exp(\alpha - \varepsilon)t_k$$

or, which is the same, the inequality

$$\chi[f] \geqslant \lim_{k\to\infty}\frac{1}{t_k}\ln|f(t_k)| \geqslant \alpha - \varepsilon$$

holds. Therefore, we have

$$\chi[f] \geqslant \alpha.$$

Thus, if the conditions (2.1.2) and (2.1.3) are satisfied simultaneously, we have

$$\chi[f] = \alpha. \quad \square$$

REMARK 2.1.1. In fact, Lemma 2.1.1 shows that a function $f(t)$ having the characteristic exponent $\alpha \neq \pm\infty$ is such that $|f(t)|$ grows slower than any $\exp(\alpha+\varepsilon)t$ as $t\to\infty$ and, along a certain sequence of values of $t$, faster than $\exp(\alpha-\varepsilon)t$.

THEOREM 2.1.1. *The characteristic exponent of the sum of a finite number of the functions $f_k(t)$, $k = 1, \ldots, n$, does not exceed the greatest of the characteristic exponents of these functions and coincides with it if only one function has the greatest exponent.*

EXAMPLES.

$$\chi[e^{2t} + e^{-1} + e^{3t}] = 3,$$
$$\chi[e^{-t} + e^{t} + (1 - e^{t})] = 0.$$

PROOF. 1. Let

$$\max_k \chi[f_k(t)] = \alpha \neq \pm\infty.$$

Consider

$$\lim_{t\to\infty}\frac{\left|\sum_{k=1}^n f_k(t)\right|}{\exp(\alpha+\varepsilon)t} \leqslant \sum_{k=1}^n \lim_{t\to\infty}\frac{|f_k(t)|}{\exp(\alpha+\varepsilon)t} = 0.$$

Thus,

$$\chi\left[\sum_{k=1}^n f_k(t)\right] \leqslant \alpha. \tag{2.1.5}$$

2. Let

$$\alpha = \chi[f_l(t)] > \chi[f_k(t)] = \alpha_k, \qquad k \neq l.$$

According to Lemma 2.1.1, there exists a sequence $t_m \to \infty$, $m \to \infty$, such that

$$\lim_{m\to\infty} \frac{|f_l(t_m)|}{\exp(\alpha - \varepsilon)t_m} = \infty.$$

For $\alpha_k \neq -\infty$, we have

$$\frac{\left|\sum_{k=1}^n f_k(t_m)\right|}{\exp(\alpha-\varepsilon)t_m} \geqslant \frac{|f_l(t_m)|}{\exp(\alpha-\varepsilon)t_m} - \sum_{k\neq l} \frac{|f_k(t_m)|}{\exp(\alpha_k+\varepsilon)t_m \exp(\alpha-\alpha_k-2\varepsilon)t_m}.$$

For $0 < \varepsilon < \min_{k\neq l}(\alpha - \alpha_k)/2$ the second term on the right-hand side of the last inequality tends to zero and the first tends to infinity as $m \to \infty$. Therefore,

$$\chi\left[\sum_{k=1}^n f_k(t)\right] \geqslant \alpha.$$

Comparing this inequality with that proved in item 1, we obtain

$$\chi\left[\sum_{k=1}^n f_k(t)\right] = \alpha. \quad \square$$

REMARK 2.1.2. It is essential in the proof that $\alpha_k \neq \pm\infty$. Formally, the inequality (2.1.5) remains valid without this requirement.

THEOREM 2.1.2. *The characteristic exponent of the product of a finite number of functions does not exceed the sum of the characteristic exponents of the cofactors,*

$$\chi\left[\prod_{k=1}^n f_k(t)\right] \leqslant \sum_{k=1}^n \chi[f_k(t)]. \tag{2.1.6}$$

REMARK 2.1.3. We assume that $+\infty$ and $-\infty$ are not simultaneously present among the characteristic exponents.

PROOF.

$$\begin{aligned}
\chi\left[\prod_{k=1}^n f_k(t)\right] &= \overline{\lim_{t\to\infty}} \frac{1}{t} \ln\left|\prod_{k=1}^n f_k(t)\right| \\
&= \overline{\lim_{t\to\infty}} \frac{1}{t} \ln \prod_{k=1}^n |f_k(t)| \\
&= \overline{\lim_{t\to\infty}} \sum_{k=1}^n \frac{1}{t} \ln|f_k(t)| \leqslant \sum_{k=1}^n \overline{\lim_{t\to\infty}} \frac{1}{t} \ln|f_k(t)| \\
&= \sum_{k=1}^n \chi[f_k],
\end{aligned}$$

which implies (2.1.6). $\square$

EXAMPLES 2.1.1. 1) $\chi[e^t e^{-3t}] = \chi[e^{-2t}] = -2$, or $\chi[e^t e^{-3t}] = \chi[e^t] + \chi[e^{-3t}] = 1 - 3 = -2$. In this case a strict equality is realized in (2.1.6).

2) $\chi[e^{t\sin t} e^{-t\sin t}] = \chi[1] = 0$. At the same time

$$\chi\left[e^{t\sin t} e^{-t\sin t}\right] < \chi\left[e^{t\sin t}\right] + \chi\left[e^{-t\sin t}\right] = 2.$$

In this case a strict inequality is realized in (2.1.6).

DEFINITION 2.1.2. The characteristic exponent of $f(t)$ is called *sharp* if there exists a finite limit

$$\lim_{t\to\infty} \frac{1}{t} \ln|f(t)| = \alpha. \tag{2.1.7}$$

THEOREM 2.1.3. *A function $f(t)$ has a sharp characteristic exponent if and only if the equality*

$$\chi[f] + \chi[1/f] = 0 \tag{2.1.8}$$

*is valid.*

REMARK 2.1.4. If follows from Definition 2.1.2 that $f(t) \neq 0$ for $t > T$.

PROOF. *Necessity.* By virtue of (2.1.7), we have

$$-\chi[f] = -\lim_{t\to\infty} \frac{1}{t} \ln|f(t)| = \lim_{t\to\infty} \frac{1}{t} \ln|1/f(t)| = \chi[1/f].$$

This implies that (2.1.8) holds.

*Sufficiency.* To prove that (2.1.7) is valid, it is sufficient to show that (2.1.8) implies the equality

$$\overline{\lim_{t\to\infty}} \frac{1}{t} \ln|f(t)| = \underline{\lim_{t\to\infty}} \frac{1}{t} \ln|f(t)|.$$

Indeed,

$$\chi[1/f] = \overline{\lim_{t\to\infty}} \frac{1}{t} \ln|1/f(t)| = -\underline{\lim_{t\to\infty}} \frac{1}{t} \ln|f(t)| \overset{(2.1.8)}{=} -\overline{\lim_{t\to\infty}} \frac{1}{t} \ln|f(t)|. \quad \square$$

REMARK 2.1.5. It is essential in Examples 2.1.1 that in the first example the functions have sharp characteristic exponents and in the second they do not.

THEOREM 2.1.4. *If a function $f(t)$ has a sharp characteristic exponent, then the characteristic exponent of the product of functions $f(t)$ and $g(t)$ is equal to the sum of their characteristic exponents, i.e.,*

$$\chi[fg] = \chi[f] + \chi[g]. \tag{2.1.9}$$

PROOF. According to Theorem 2.1.2,

$$\chi[fg] \leqslant \chi[f] + \chi[g].$$

At the same time we have

$$\chi[g] = \chi[gf(1/f)] \leqslant \chi[fg] - \chi[f],$$

or

$$\chi[g] + \chi[f] \leqslant \chi[fg].$$

This inequality together with the first one gives (2.1.9). $\square$

COROLLARY 2.1.1. $\chi[e^{\alpha t} f(t)] = \alpha + \chi[f]$.

Now let us consider the characteristic exponent of an integral. Naturally, it should be related to the characteristic exponent of the integrand. Let $F(t) = \int_{t_0}^{t} e^{\alpha\tau}\, d\tau$.

$$\begin{array}{llll}
\text{For} \quad \alpha > 0, & F(t) = \dfrac{1}{\alpha}(e^{\alpha t} - e^{\alpha t_0}) & \Rightarrow & \chi[F] = \alpha. \\
\text{For} \quad \alpha = 0, & F(t) = t - t_0 & \Rightarrow & \chi[F] = \alpha. \\
\text{For} \quad \alpha < 0, & F(t) = \dfrac{1}{\alpha}(e^{\alpha t} - e^{\alpha t_0}) & \Rightarrow & \chi[F] = 0.
\end{array}$$

In the last case there is an inconsistency. If we set

$$F(t) = \int_{\infty}^{t} e^{\alpha\tau}\, d\tau = \frac{1}{\alpha} e^{\alpha t} \qquad \text{for} \qquad \alpha < 0,$$

then $\chi[F] = \alpha$.

Following Lyapunov **[25]**, when considering characteristic exponents of integrals, we set

$$F(t) = \int_{a}^{t} f(\tau)\, d\tau, \qquad \text{where} \qquad a = \begin{cases} t_0, & \text{for} \quad \chi[f] \geqslant 0, \\ \infty, & \text{for} \quad \chi[f] < 0. \end{cases}$$

The integral defined in this way is called the *Lyapunov integral.*

THEOREM 2.1.5. *The characteristic exponent of an integral does not exceed the characteristic exponent of the integrand.*

PROOF. Let $\chi[f] = \alpha \neq \pm\infty$. By Lemma 2.1.1, for any $\varepsilon > 0$ we have

$$|f(t)| \leqslant M e^{(\alpha+\varepsilon)t}, \qquad t \geqslant t_0,$$

where $M$ is some constant depending, generally speaking, on $\varepsilon$.

1. Let $\alpha \geqslant 0$; then

$$\begin{aligned} |F(t)| &\leqslant \int_{t_0}^{t} |f(\tau)|\, d\tau \leqslant \int_{t_0}^{t} M e^{(\alpha+\varepsilon)\tau}\, d\tau \\ &= \frac{M}{\alpha+\varepsilon}\left[e^{(\alpha+\varepsilon)t} - e^{(\alpha+\varepsilon)t_0}\right] < \frac{M}{\alpha+\varepsilon} e^{(\alpha+\varepsilon)t}. \end{aligned}$$

By the monotonicity of the characteristic exponent, we have $\chi[F] \leqslant \alpha + \varepsilon$; since $\varepsilon$ is arbitrary, this implies that

$$\chi[F] \leqslant \alpha = \chi[f].$$

2. Now let $\alpha < 0$ and $0 < \varepsilon < |\alpha|$; then

$$|F(t)| \leqslant \int_{t}^{\infty} |f(\tau)|\, d\tau \leqslant \int_{t}^{\infty} M e^{(\alpha+\varepsilon)\tau}\, d\tau = \frac{M e^{(\alpha+\varepsilon)t}}{|\alpha+\varepsilon|}.$$

In analogy to the previous, we have

$$\chi[F] \leqslant \alpha = \chi[f].$$

3. The statement of the theorem still holds for $\alpha = +\infty$ or $\alpha = -\infty$. □

## §2. Characteristic exponents of matrices of functions

Let us consider a matrix

$$F(t) = \{f_{ij}(t)\}, \qquad i = 1, \dots, n, \quad j = 1, \dots, m, \quad m \leqslant n,$$

defined for $t \in [t_0, \infty)$.

DEFINITION 2.2.1. The number (or the symbol $\pm\infty$) defined as

$$\chi[F] = \max_{ij} \chi[f_{ij}]$$

is called the *characteristic exponent of the matrix* $F(t)$.

Note that $\chi[F] = \chi[F^*]$, which follows from the definition.

LEMMA 2.2.1. *The characteristic exponent of a finite-dimensional matrix $F(t)$ coincides with the characteristic exponent of its norm, i.e.,*

$$\chi[F] = \chi[\|F\|].$$

PROOF. We shall carry out the proof for the three norms introduced in Chapter I, §2. The estimate

$$|f_{ij}(t)| \leqslant \|F(t)\|, \qquad t \in [t_0, \infty), \quad i = 1, \dots, n, \quad j = 1, \dots, m,$$

is valid for any of these norms. This is obvious for $\|F(t)\|_I$ and $\|F(t)\|_{II}$, and for the third norm there is the estimate

$$\max_i \left( \sum_{j=1}^{m} |f_{ij}(t)|^2 \right)^{1/2} \leqslant \|F(t)\|_{III}.$$

From these two inequalities, by the monotonicity of the characteristic exponent, we have

$$\chi[F] \leqslant \chi[\|F\|]. \tag{2.2.1}$$

At the same time

$$\|F(t)\| \leqslant \sum_{i,j} |f_{ij}(t)|,$$

which is obvious for the first two norms, and for the third norm we use the inequality

$$\|F(t)\|_{III} \leqslant \left( \sum_{ij} |f_{ij}|^2 \right)^{1/2}.$$

It follows from the last two inequalities that

$$\chi[\|F\|] \leqslant \chi[F]. \tag{2.2.2}$$

Comparing (2.2.1) with (2.2.2), we obtain what was required. □

COROLLARY 2.2.1. *Let $x(t)$ be a vector-function defined for $t \in [t_0, \infty)$; then*

$$\chi[x] - \overline{\lim_{t\to\infty}} \frac{1}{t} \ln \|x(t)\|.$$

THEOREM 2.2.1. *The characteristic exponent of the sum of a finite number of matrices does not exceed the greatest characteristic exponent and is equal to it if only one matrix has the greatest characteristic exponent.*

PROOF. Let $F(t) = \sum_{k=1}^{N} F_k(t)$; then

$$\|F(t)\| \leqslant \sum_{k=1}^{N} \|F_k(t)\|.$$

By the monotonicity of the characteristic exponent and Theorem 2.1.1, we have

$$\chi[\|F(t)\|] \leqslant \chi\left[\sum_{k=1}^{N} \|F_k(t)\|\right] \leqslant \max_k \chi[\|F_k(t)\|] = \max_k \chi[F_k(t)].$$

The first part of the theorem is proved.

Now let $\chi[F_1] > \chi[F_k]$, $k > 1$, and let this characteristic exponent be realized by the element $f_{pq}^{(1)}(t)$ of the matrix $F_1(t)$. Here $F(t) = \{f_{ij}(t)\}$ and $F_k(t) = \{f_{ij}^{(k)}(t)\}$. By the addition rule for matrices,

$$f_{pq}(t) = \sum_{k=1}^{N} f_{pq}^{(k)}(t),$$

and, by Theorem 2.1.1, we have

$$\chi[f_{pq}] = \chi[f_{pq}^{(1)}].$$

At the same time,

$$\chi[F] \geqslant \chi[f_{pq}] = \chi[F_1] = \max_k \chi[F_k].$$

Comparing this inequality with the result of the first part of the theorem, we obtain the equality required,

$$\chi[F] = \max_k \chi[F_k]. \quad \square$$

THEOREM 2.2.2. *The characteristic exponent of the product of a finite number of matrices does not exceed the sum of the characteristic exponents of these matrices.*

PROOF. Let $F(t) = \prod_{s=1}^{N} F_s(t)$; therefore, $\|F(t)\| \leqslant \prod_{s=1}^{N} \|F_s(t)\|$. Using Theorem 2.1.2, we obtain the inequality required,

$$\chi[F] = \chi[\|F\|] \leqslant \sum_{s=1}^{N} \chi[\|F_s\|] = \sum_{s=1}^{N} \chi[F_s]. \quad \square$$

COROLLARY 2.2.2. *The characteristic exponent of the linear combination*

$$\sum_{k=1}^{N} c_k F_k(t)$$

*of a finite number of matrices does not exceed the greatest characteristic exponent of these matrices and coincides with it if only one matrix has the greatest characteristic exponent.*

## §3. The spectrum of a linear system

First let us prove a theorem of a more general character.

THEOREM 2.3.1. *Any nontrivial solution of the normal system*

$$\dot{x} = f(t,x), \qquad x \in \mathbb{C}^n, \qquad \|f(t,x)\| \leqslant L\|x\|,$$

*has a finite characteristic exponent.*

REMARK 2.3.1. By the condition imposed on the right-hand side of the system, any solution of the system is defined for all $t \in \mathbb{R}$.

PROOF. Let $\|x\| = (x,x)^{1/2}$ and consider a nontrivial solution $x(t)$. We have

$$\begin{aligned} \left|d\|x\|^2/dt\right| &= |(\dot{x},x) + (x,\dot{x})| = 2|\operatorname{Re}(\dot{x},x)| \\ &= 2|\operatorname{Re}(f(t,x),x)| \leqslant 2L\|x\|^2. \end{aligned}$$

Therefore,

$$-2L \leqslant \frac{d\|x\|^2/dt}{\|x\|^2} \leqslant 2L.$$

Let us integrate the last inequality from $t_0$ to $t$:

$$-2L(t-t_0) \leqslant 2\ln\|x(t)\| - 2\ln\|x(t_0)\| \leqslant 2L(t-t_0).$$

Dividing by $t$ and letting $t \to \infty$, we obtain

$$-L \leqslant \chi[\|x\|] \leqslant L. \quad \square$$

Now we pass to linear systems

$$\dot{x} = A(t)x, \qquad x \in \mathbb{C}^n, \quad A \in C[t_0,\infty). \tag{2.3.1}$$

THEOREM 2.3.1′. *If* $\sup_t \|A(t)\| \leqslant M$, *then any nontrivial solution* $x(t)$ *of the linear system* (2.3.1) *has a finite characteristic exponent, and* $-M \leqslant \chi[x] \leqslant M$.

PROOF. The right-hand side of the linear system satisfies the condition of Theorem 2.3.1 with the constant $L = M$. Thus, by Theorem 2.3.1, we have what was required.□

REMARK 2.3.2. Basic facts of the theory of linear systems were obtained under the assumption that the coefficients of the systems are bounded. We shall also impose this assumption, paying attention to every possibility to slacken it.

REMARK 2.3.3. If the matrix $A(t)$ is real and there exists a complex solution $z(t)$ having the characteristic exponent $\alpha$, then there exists a real solution with the same characteristic exponent.

PROOF. Let $z(t) = z_1(t) + iz_2(t)$. Recall that $z_1(t)$ and $z_2(t)$ are solutions; this follows from $A(t)$ being real. Let us consider two cases.

1. $\chi[z_1] \neq \chi[z_2]$. Let, for example, $\chi[z_1] > \chi[z_2]$.
By Theorem 2.1.1, $\chi[z] = \chi[z_1]$, i.e., $z_1(t)$ is the required one.
2. $\chi[z_1] = \chi[z_2] = \beta$. By Theorem 2.1.1, $\alpha \leqslant \beta$.
At the same time, $z_1(t) = \frac{1}{2}[z(t) - \bar{z}(t)]$, which implies that $\beta \leqslant \alpha$ by the same theorem, i.e., $\alpha = \beta$ and $z_1(t)$ is again the solution in question. □

THEOREM 2.3.2. *Vector-functions $x_1(t), \dots, x_m(t)$ defined on $[t_0, \infty)$ and having different finite characteristic exponents are linearly independent.*

PROOF. Let us consider the linear combination of these vectors with a nontrivial set of coefficients. According to Theorem 2.1.1, the characteristic exponent of this sum is equal to $\max_i \chi[x_i] \neq -\infty$; hence, this linear combination cannot be an identical zero. □

COROLLARY 2.3.1. *The solutions of a linear system cannot have more than n different characteristic exponents.*

DEFINITION 2.3.1. The set of all different characteristic exponents of solutions of the linear system is called the *spectrum*. The number of elements of the spectrum does not exceed $n$.

REMARK 2.3.4. A nonlinear system can have an infinite spectrum:

$$t\dot{x} = x \ln x \quad \Rightarrow \quad x = \exp ct.$$

LEMMA 2.3.1. *The characteristic exponent of a solution can be calculated using sequences of integers, i.e., the formula*

$$\chi[x] = \overline{\lim_{n\to\infty}} \frac{1}{n} \ln \|x(n)\|, \tag{2.3.2}$$

*where n is a natural number, is valid.*

PROOF. Obviously,

$$\overline{\lim_{n\to\infty}} \frac{1}{n} \ln \|x(n)\| \leqslant \overline{\lim_{t\to\infty}} \frac{1}{t} \ln \|x(t)\| = \chi[x].$$

If we indicate a sequence of integers $n_k \underset{k\to\infty}{\longrightarrow} \infty$ such that

$$\lim_{k\to\infty} \frac{1}{n_k} \ln \|x(n_k)\| \geqslant \chi[x], \tag{2.3.3}$$

then the lemma will be proven since this sequence realizes equality (2.3.2).

Let $t_k \underset{k\to\infty}{\longrightarrow} \infty$ be a sequence realizing the upper limit in the definition of the characteristic exponent, i.e.,

$$\lim_{k\to\infty} \frac{1}{t_k} \ln \|x(t_k)\| = \chi[x].$$

Let us set $n_k = [t_k]$ and show that this sequence is the required one. Indeed,

$$\begin{aligned} r_k &= \frac{1}{t_k} \ln \|x(t_k)\| - \frac{1}{n_k} \ln \|x(n_k)\| \\ &= \left( \frac{\ln \|x(t)\|}{t} \right)'_{t=\xi\in(n_k,t_k)} (n_k - t_k) \\ &= \left( \frac{\|x(t)\|'}{\|x(t)\| t} - \frac{\ln \|x(t)\|}{t^2} \right)_{t=\xi} (n_k - t_k). \end{aligned}$$

Let us estimate the value of $r_k$, using the following facts:

a) $\left| \|x\| \frac{d\|x(t)\|}{dt} \right| \leqslant M\|x\|^2$ (see Theorem 2.3.1),

b) for sufficiently large $t$, the inequality

$$\left|\frac{\ln \|x(t)\|}{t}\right| \leqslant 2M$$

holds.

Thus,

$$|r_k| \leqslant \left(\left|\frac{\|x(t)\| \cdot \|x(t)\|'}{\|x(t)\|^2 t}\right| + \left|\frac{\ln \|x(t)\|}{t} \cdot \frac{1}{t}\right|\right)_{t=\xi} \leqslant \frac{M}{n_k} + \frac{2M}{n_k} = \frac{3M}{n_k}.$$

Therefore, we have

$$-\frac{3M}{n_k} + \frac{1}{t_k} \ln \|x(t_k)\| \leqslant \frac{1}{n_k} \ln \|x(n_k)\| \leqslant \frac{1}{t_k} \ln \|x(t_k)\| + \frac{3M}{n_k}.$$

By the squeeze convergence principle, $\lim_{k\to\infty} \frac{1}{n_k} \ln \|x(n_k)\|$ exists, and from the left-hand side of the last inequality we have the inequality (2.3.3) in question. □

## §4. Normal fundamental systems or normal bases

Let us turn again to the linear homogeneous system (2.3.1). Let its spectrum contain $m \leqslant n$ elements, i.e., let

$$-\infty < \alpha_1 < \alpha_2 < \cdots < \alpha_m < \infty$$

be the set of different characteristic exponents of this system.

Consider some fundamental system of solutions $X(t) = \{x_1(t), \dots, x_n(t)\}$ ordered in such a way that $\chi[x_k] \leqslant \chi[x_{k+1}]$, $k = 1, \dots, n-1$. However, in this order there is an arbitrariness in the indexing of vectors with identical characteristic exponents. Let this system contain $r_k$ solutions with the characteristic exponent $\alpha_k$, $k = 1, \dots, m$. We note that some $r_k$ may be zero.

EXAMPLE 2.4.1. Consider the system

$$\dot{x} = \operatorname{diag}[1, 2, 3]x$$

and its two fundamental systems of solutions

$$X_1(t) = \{(e^t, 0, 0)^\top, (0, e^{2t}, 0)^\top, (0, 0, e^{3t})^\top\},$$
$$X_2(t) = \{(e^t, 0, e^{3t})^\top, (0, e^{2t}, e^{3t})^\top, (0, 0, e^{3t})^\top\}.$$

The last fundamental system does not have solutions with the characteristic exponents 1 and 2 at all.

DEFINITION 2.4.1. The number

$$\sigma_X = \sum_{k=1}^{m} r_k \alpha_k$$

is called the *sum of characteristic exponents of the fundamental system* $X(t)$.

In Example 2.4.1, $\sigma_{X_1} = 6$, $\sigma_{X_2} = 9$. Since the spectrum of the system is finite and the number of linearly independent solutions is equal to $n$, the value $\sigma_X$ attains its minimum on the set of fundamental systems.

DEFINITION 2.4.2. A fundamental system $X(t)$ is called *normal* if the sum of its characteristic exponents is minimal in comparison with other fundamental systems.

In order to give a more complete characterization of a normal fundamental system, Lyapunov **[25]** introduced the notion of incompressibility.

DEFINITION 2.4.2′. A set of nontrivial vector-functions $x_1(t), \dots, x_k(t)$ has the property of *incompressibility* if the characteristic exponent of any linear combination with a nontrivial set of coefficients is equal to the greatest characteristic exponent of the vector-functions entering the combination.

Note that, for example, a set of vector-functions with different characteristic exponents is incompressible (see Theorem 2.1.1). From the incompressibility of vector-functions with finite characteristic exponents it follows that they are linearly independent. Indeed, otherwise there would exist a linear combination identically equal to zero whose characteristic exponent is equal to $-\infty$. The converse statement is not true. Thus, in Example 2.4.1 the set of linearly independent solutions $X_2(t)$ is compressible.

Denote by $N_s$ the greatest possible number of linearly independent solutions of system (2.3.1) having the characteristic exponent $\alpha_s$, $s = 1, \dots, m$. Note that this number is inherent to the initial system (2.3.1) itself and is not connected in any way with the choice of the fundamental system $X(t)$. If the latter has $r_s$ solutions with the characteristic exponent $\alpha_s$, then it is obvious that

$$N_s \geqslant r_1 + r_2 + \cdots + r_s, \tag{2.4.1}$$

since the solutions with smaller characteristic exponents can be absorbed by solutions with greater ones (see Example 2.4.1).

LEMMA 2.4.1. *If a fundamental system is incompressible, then*

$$N_s = r_1 + r_2 + \cdots + r_s, \qquad s = 1, \dots, m. \tag{2.4.2}$$

PROOF. It follows from the incompressibility that any solution with the characteristic exponent $\alpha_s$ can be a linear combination of only those elements of the fundamental system whose exponents do not exceed $\alpha_s$, i.e.,

$$N_s \leqslant r_1 + r_2 + \cdots + r_s, \qquad s = 1, \dots, m. \tag{2.4.3}$$

Comparing inequalities (2.4.1) and (2.4.3), we obtain the equality (2.4.2) required. □

THEOREM 2.4.1 (Lyapunov). *A fundamental system of solutions is normal if and only if it is incompressible.*

PROOF. *Necessity.* Let $X(t) = \{x_1(t), \dots, x_n(t)\}$ be a normal fundamental system. Let us show that it is incompressible. Let us assume that $\chi[x_k] \leqslant \chi[x_{k+1}]$, $k = 1, \dots, n-1$, and, still, let there exist a reducing linear combination

$$y = \sum_{j=1}^{l} c_j x_j(t), \qquad |c_l| \neq 0, \qquad \chi[y] < \chi[x_l], \qquad l \leqslant n.$$

Consider the set of vectors

$$Y(t) = \{x_1(t), \dots, x_{l-1}(t), y(t), x_{l+1}(t), \dots, x_n(t)\}.$$

They are linearly independent (since, otherwise, the set $X(t)$ would be linearly dependent), each of them is a solution and, thus, $Y(t)$ is a fundamental system; moreover, $\sigma_Y < \sigma_X$. This contradicts the fact that the system $X(t)$ is normal.

*Sufficiency.* Let a fundamental system $X(t) = \{x_1(t), \dots, x_n(t)\}$ contain $n_s$ solutions with the characteristic exponent $\alpha_s$, $s = 1, \dots, m$, and be incompressible. Let us show that it is normal. And yet, let there exist a fundamental system $Y(t) = \{y_1(t), \dots y_n(t)\}$ having $r_s$ solutions with the characteristic exponent $\alpha_s$ such that

$$\sigma_Y = \sum_{s=1}^{m} r_s\alpha_s < \sum_{s=1}^{m} n_s\alpha_s = \sigma_X. \tag{2.4.4}$$

According to Lemma 2.4.1, $N_s = n_1 + \dots + n_s$ and, moreover, $N_s \geqslant r_1 + \dots + r_s = N_s'$, $s = 1, \dots, m$. Let us set $N_0' = N_0 = 0$ and consider

$$\begin{aligned}
\sigma_Y = \sum_{s=1}^{m} r_s\alpha_s &= \sum_{s=1}^{m} (N_s' - N_{s-1}')\alpha_s \\
&= N_m'\alpha_m - \sum_{s=1}^{m-1} N_s'(\alpha_{s+1} - \alpha_s) \\
&\geqslant N_m\alpha_m - \sum_{s=1}^{m-1} N_s(\alpha_{s+1} - \alpha_s) \\
&= \sum_{s=1}^{m} n_s\alpha_s = \sigma_X.
\end{aligned}$$

The inequality obtained contradicts (2.4.4). □

Now let us discuss the structure of the set of solutions of the linear system (2.3.1). This set forms a linear space $L^n$ of dimension $n$. A fundamental system $X(t) = \{x_1(t), \dots, x_n(t)\}$ is a basis of this space. Any $l$-dimensional linear subspace $L^l$ of the space $L^n$ will be called a *lineal*.

Let $\alpha_s$ be an element of the spectrum of the system. Consider the set $L$ of all the solutions $x(t)$, including the trivial one, whose characteristic exponents do not exceed $\alpha_s$. Let us show that this set is a lineal $L^{N_s}$. Recall that $N_s$ is the greatest number of linearly independent solutions of the system having the characteristic exponent $\alpha_s$. We verify that the dimension of the lineal $L$ under consideration is indeed $N_s$. On one hand,

$$\dim L \geqslant N_s, \tag{2.4.5}$$

since all the solutions with characteristic exponent $\alpha_s$ belong to $L$. On the other hand, a basis of this subspace necessarily contains at least one solution with characteristic exponent $\alpha_s$. Adding this solution to the other basis solutions with smaller characteristic exponents, we obtain a new basis all of whose elements have the characteristic exponent $\alpha_s$, i.e.,

$$\dim L \leqslant N_s. \tag{2.4.6}$$

From the inequalities (2.4.5) and (2.4.6) it follows that

$$\dim L = N_s.$$

according to Theorem 2.1.1, this sum has the greatest characteristic exponent. Hence, the basis

$$X(t) = \{x_1(t), \dots, x_n(t)\}$$

is incompressible and, therefore, normal. □

## §5. The Lyapunov inequality for the sum of characteristic exponents of bases

Let $X(t) = \{x_1(t), \dots, x_n(t)\}$ be a basis or, what is the same, a fundamental system of solutions of system (2.3.1), and let $\sigma_X$ be the sum of its characteristic exponents (see Definition 2.4.1).

THEOREM 2.5.1 (Lyapunov). *For any fundamental system $X(t)$ the inequality*

$$\sigma_X \geqslant \chi\left[e^{\int_{t_0}^{t} \operatorname{Sp} A(\tau)\, d\tau}\right] = \overline{\lim_{t\to\infty}} \frac{1}{t} \int_{t_0}^{t} \operatorname{Re} \operatorname{Sp} A(u)\, du \tag{2.5.1}$$

*is valid.*

PROOF. According to the Ostrogradskiĭ-Liouville formula, we have

$$\operatorname{Det} X(t) = \operatorname{Det} X(t_0) \exp \int_{t_0}^{t} \operatorname{Sp} A(\tau)\, d\tau.$$

Thus,

$$\chi[\operatorname{Det} X] = \chi\left[\exp \int_{t_0}^{t} \operatorname{Sp} A(\tau)\, d\tau\right].$$

The determinant is the sum of $n!$ summands each of which is the product of $n$ elements of the matrix from different columns. From this and Theorems 2.1.1 and 2.1.2 it follows that

$$\chi[\operatorname{Det} X] \leqslant \sum_{s=1}^{m} r_s \alpha_s = \sigma_X.$$

This implies that (2.5.1) holds. □

REMARK 2.5.1. The inequality (2.5.1) is called the *Lyapunov inequality* for the sum of characteristic exponents of a fundamental system or a basis.

COROLLARY 2.5.1. *If for some basis the Lyapunov equality holds, then this basis is normal.*

REMARK 2.5.2. There exist normal bases for which the Lyapunov equality does not hold.

We illustrate this statement by the following example.

EXAMPLE 2.5.1.

$$\begin{aligned} \dot{x} &= y[\sin \ln t + \cos \ln t], \\ \dot{y} &= x[\sin \ln t + \cos \ln t], \end{aligned} \qquad t \geqslant 1,$$

$$X(t) = \begin{pmatrix} e^{t \sin \ln t} & e^{-t \sin \ln t} \\ e^{t \sin \ln t} & -e^{-t \sin \ln t} \end{pmatrix} = \{x_1(t), x_2(t)\}.$$

This basis is incompressible. Indeed,

$$\chi[c_1 e^{t \sin \ln t} + c_2 e^{-t \sin \ln t}] \leqslant 1.$$

But the sequence $t_k = \exp[\pi/2 + 2k\pi]$ realizes the exponent equal to unity:

$$\begin{aligned}
\lim_{k\to\infty} \frac{1}{t_k} \ln \left| c_1 e^{t_k \sin \ln t_k} + c_2 e^{-t_k \sin \ln t_k} \right| \\
&= \lim_{k\to\infty} \frac{1}{t_k} \ln \left( e^{t_k \sin \ln t_k} \left| c_1 + c_2 e^{-2t_k \sin \ln t_k} \right| \right) \\
&= 1.
\end{aligned}$$

Thus, $X(t)$ is a normal basis and $\chi[x_1] = \chi[x_2] = 1$, i.e., $\sigma_X = 2$. At the same time,

$$\chi\left[e^{\int_0^t \operatorname{Sp} A(\tau)\,d\tau}\right] = \chi[e^0] = 0.$$

Hence, the strict Lyapunov inequality holds for a normal basis of this system.

CHAPTER III

# Reducible, Almost Reducible, and Regular Systems

Among linear systems, those with constant coefficients are the simplest and are completely understood. They can be solved explicitly. The next most thoroughly studied class comprises linear systems with periodic coefficients. We know a lot about these systems, but they cannot be solved explicitly. Lyapunov [**25**] showed that there exists a transformation which does not change the character of the growth of solutions and reduces a system with periodic coefficients to a system with constant coefficients. The systems having this property were called reducible by Lyapunov, and the indicated transformation is called the Lyapunov transformation. The theory of reducible systems was advanced with the publication of Erugin [**20**], which is interesting from the point of view of its results as well as methodology.

Afterwards, the notion of reducibility was extended, and there appeared a notion of reducibility of one system to another if their matrix coefficients are connected by a Lyapunov transformation. It is not assumed any longer that any of these systems is autonomous. The fact is that Lyapunov transformations, preserving the main characteristics of a system, often simplify it, which is very useful. In 1962 Bylov [**10**] introduced the notion of almost reducibility, i.e., reducibility with a small error.

We distinguish five classes in the set of linear systems:

1) systems with constant coefficients,
2) systems with periodic coefficients,
3) systems reducible to autonomous ones,
4) systems almost reducible to autonomous ones,
5) regular systems.

Each subsequent class includes the preceding ones. Chapter I was dedicated to the first two classes. In Chapter III we study the properties of reducibility and almost reducibility. The second part of this chapter is dedicated to regular systems defined by Lyapunov [**25**]; they play an important role in the stability theory in linear approximation.

Now we proceed with a systematic exposition.

## §1. Lyapunov transformations

Consider a system

$$\dot{x} = A(t)x, \tag{3.1.1}$$

where

$$x \in \mathbb{C}^n, \qquad A \in C[t_0, \infty), \qquad \sup_{t \geqslant t_0} \|A(t)\| \leqslant M,$$

and a transformation

$$x = L(t)y \tag{3.1.2}$$

with a nonsingular and continuously differentiable for $t \geqslant t_0$ matrix $L(t)$. Applying (3.1.2) to system (3.1.1) we again obtain a linear system

$$\dot{y} = B(t)y, \tag{3.1.3}$$

$$B(t) = L^{-1}(t)A(t)L(t) - L^{-1}(t)\dot{L}(t). \tag{3.1.4}$$

The relation (3.1.4) between the matrices $A(t)$ and $B(t)$ is called *kinematic similarity* and in the case when $L(t) \equiv \text{const}$, *static similarity*. If we wish that system (3.1.3) remain in the class of systems with bounded coefficients it is necessary to strengthen the conditions on the matrix $L(t)$ and require that the matrices $L(t)$, $L^{-1}(t)$, and $\dot{L}(t)$ be bounded for $t \geqslant t_0$.

DEFINITION 3.1.1. The transformation (3.1.2) is called *the Lyapunov transformation* if

1) $L \in C^1[t_0, \infty)$,
2) $L(t), L^{-1}(t), \dot{L}(t)$ are bounded for $t \geqslant t_0$.

The matrix $L(t)$ having these properties is said to be a *Lyapunov matrix*.

We note several properties of Lyapunov transformations.

1. *Lyapunov transformations form a group.*

Indeed,

a) if $L(t)$ is a Lyapunov matrix, then $L^{-1}(t)$ is also a Lyapunov matrix (this statement immediately follows from the construction of the inverse matrix),
b) $L(t) \equiv E$ is a Lyapunov matrix (this is verified by checking the properties from Definition 3.1.1),
c) if $L_1(t)$ and $L_2(t)$ are Lyapunov matrices, then $L(t) = L_1(t)L_2(t)$ is a Lyapunov matrix (verified analogously).

2. *Lyapunov transformations do not change characteristic exponents.* Let $x = L(t)y$ and $\|L(t)\| \leqslant K$, $\|L^{-1}(t)\| \leqslant K$ for $t \geqslant t_0$. Obviously,

$$\|x\| \leqslant \|L(t)\|\|y\| \leqslant K\|y\| \quad \Rightarrow \quad \chi[x] \leqslant \chi[y].$$

At the same time, $y = L^{-1}(t)x$; therefore

$$\|y\| \leqslant K\|x\| \quad \Rightarrow \quad \chi[y] \leqslant \chi[x],$$

and finally we have

$$\chi[x] = \chi[y].$$

It follows from the facts proved above that the complete spectra of systems (3.1.1) and (3.1.3) coincide.

3.

$$\overline{\lim_{t\to\infty}} \frac{1}{t} \int_{t_0}^{t} \operatorname{Re} \operatorname{Sp} A(\tau)\, d\tau = \overline{\lim_{t\to\infty}} \frac{1}{t} \int_{t_0}^{t} \operatorname{Re} \operatorname{Sp} B(\tau)\, d\tau,$$

$$\underline{\lim_{t\to\infty}} \frac{1}{t} \int_{t_0}^{t} \operatorname{Re} \operatorname{Sp} A(\tau)\, d\tau = \underline{\lim_{t\to\infty}} \frac{1}{t} \int_{t_0}^{t} \operatorname{Re} \operatorname{Sp} B(\tau)\, d\tau.$$

Indeed, the fundamental matrices $X(t)$ and $Y(t)$ of systems (3.1.1) and (3.1.3) are connected by the relation

$$X(t) = L(t)Y(t).$$

Thus,

$$\operatorname{Det} X(t) = \operatorname{Det} L(t) \operatorname{Det} Y(t),$$

or

$$\left| \operatorname{Det} X(t_0) e^{\int_{t_0}^{t} \operatorname{Sp} A(\tau)\, d\tau} \right| = \left| \operatorname{Det} L(t) \operatorname{Det} Y(t_0) e^{\int_{t_0}^{t} \operatorname{Sp} B(\tau)\, d\tau} \right|.$$

Taking logarithms in the last identity, dividing by $t$, and considering the corresponding limits, we obtain the required result.

## §2. Reducible systems

DEFINITION 3.2.1. A system $\dot{x} = A(t)x$ is said to be *reducible* to a system $\dot{y} = B(t)y$, if there exists a Lyapunov transformation $x = L(t)y$ such that

$$B(t) = L^{-1}(t)A(t)L(t) - L^{-1}(t)\dot{L}(t).$$

Note the following properties of reducibility.

1. *Transitivity*. If a system $\dot{x} = A(t)x$ is reducible to $\dot{y} = B(t)y$ and the latter is reducible to a system $\dot{z} = C(t)z$, then the system $\dot{x} = A(t)x$ is reducible to $\dot{z} = C(t)z$.

Let the first reduction be realized by a transformation $x = L_1(t)y$ and the second, by a transformation $y = L_2(t)z$; then the system $\dot{x} = A(t)x$ by means of the transformation $x = L_1(t)L_2(t)z$ is reduced to the system $\dot{z} = C(t)z$.

2. *Symmetry*. If a system $\dot{x} = A(t)x$ is reducible to a system $\dot{y} = B(t)y$, then $\dot{y} = B(t)y$ is also reducible to $\dot{x} = A(t)x$ (this follows from the fact that, together with $L(t)$, the matrix $L^{-1}(t)$ is also Lyapunov).

3. If a system $\dot{x} = A(t)x$ is reduced to a system $\dot{y} = B(t)y$ by a transformation $x = L(t)y$, then the system $\dot{x} = -A^*(t)$ is reduced to $\dot{y} = -B^*(t)y$ by the transformation $x = [L^{-1}(t)]^*y$.

This statement is verified straightforwardly. If $B = L^{-1}AL - L^{-1}\dot{L}$, then

$$L^*(-A^*)(L^{-1})^* - L^*\frac{d[L^{-1}]^*}{dt} = -(L^{-1}AL - L^{-1}\dot{L})^* = -B^*.$$

DEFINITION 3.2.2. If for a system $\dot{x} = A(t)x$ there exists a Lyapunov transformation $x = L(t)y$ such that the matrix

$$B = L^{-1}(t)A(t)L(t) - L^{-1}(t)\dot{L}(t)$$

is constant, then the system is said to be *reducible to a system with constant coefficients*.

Note that originally the notion of reducibility coincided with that in Definition 3.2.2. We give a criterion of reducibility of this sort.

THEOREM 3.2.1 (Erugin). *A system $\dot{x} = A(t)x$ is reducible to a system with constant coefficients if and only if it has a fundamental matrix $X(t)$ of the form*

$$X(t) = L(t)e^{Bt}, \tag{3.2.1}$$

*where $L(t)$ is Lyapunov and $B$ is a constant matrix.*

PROOF. *Necessity.* Let the transformation $x = L(t)y$ reduce the system $\dot{x} = A(t)x$ to the system $\dot{y} = By$ with a constant matrix $B$. The complete set of fundamental matrices of the latter system has the form $Y = \exp(Bt)C$, $\operatorname{Det} C \neq 0$. Then all the fundamental matrices of the initial system have the form $X(t) = L(t)\exp(Bt)C$. For $C = E$ we obtain (3.2.1).

*Sufficiency.* Let some fundamental matrix have the form (3.2.1). Consider the transformation $x = X(t)\exp(-Bt)y$ generated by (3.2.1) and find, according to equality (3.1.4), the coefficient matrix of the system obtained by means of this transformation:

$$e^{Bt}X^{-1}(t)A(t)X(t)e^{-Bt} - e^{Bt}X^{-1}(t)[\dot{X}(t)e^{-Bt} - X(t)e^{-Bt}B] = B.$$

Here it is taken into account that $\dot{X}(t) = A(t)X(t)$. □

REMARK 3.2.1. If one of the fundamental matrices $X(t)$ of system (3.1.1) can be represented in the form (3.2.1), then any other $X_1(t)$ has an analogous form.

For the fundamental matrix $X_1(t)$ there exists a nonsingular matrix $C$ such that $X_1(t) = X(t)C$. Therefore,

$$\begin{aligned} X_1(t) &= L(t)e^{Bt}C \\ &= L(t)CC^{-1}e^{Bt}C \\ &= L_1(t)e^{B_1 t}, \end{aligned}$$

where

$$L_1(t) = L(t)C, \qquad B_1 = C^{-1}BC.$$

REMARK 3.2.2. The transformation $x = L_1(t)y$ reduces the initial system to the system $\dot{y} = B_1 y$.

The validity of this statement follows from the proof of sufficiency.

THEOREM 3.2.2 (Lyapunov). *A linear system with periodic coefficients is reducible to a system with constant coefficients.*

PROOF. By Floquet's Theorem 1.4.3 and Remark 1.4.5, any fundamental matrix $X(t)$ of a system with periodic coefficients can be represented in the form

$$X(t) = \Phi(t)e^{\Lambda t}, \qquad \Phi(t+\omega) = \Phi(t), \qquad \Lambda = \text{const}.$$

Let us show that $\Phi(t)$ is a Lyapunov matrix. Indeed, $\Phi \in C^1(R)$, and $\Phi(t)$ and $\dot{\Phi}(t)$ are bounded as periodic matrices. The boundedness of the matrix $\Phi^{-1}(t)$ follows from the inequality

$$\operatorname{Det}\Phi(t) \geqslant \min_{0\leqslant t\leqslant\omega} |\operatorname{Det}\Phi(t)| = \alpha \neq 0.$$

Thus, the matrix $X(t)$ has the form (3.2.1) and Erugin's Theorem 3.2.1 implies reducibility to a system with constant coefficients. □

Now we note one particular case. It deals with reducibility to the system with zero matrix. This situation deserves special attention because, in particular, the condition for such reducibility can be given in terms of the coefficients of the system.

THEOREM 3.2.3. *Let system* (3.1.1) *be such that*

1) *all the solutions of the system are bounded for* $t \geqslant t_0$,

$$\text{(3.2.2)} \qquad 2) \quad \int_{t_0}^{t} \operatorname{Re} \operatorname{Sp} A(\tau)\, d\tau \geqslant a > -\infty \qquad \text{for } t \geqslant t_0.$$

*Then system* (3.1.1) *is reducible to the system with zero matrix.*

PROOF. Let us show that any fundamental matrix $X(t)$ of the system is a Lyapunov matrix. Indeed,

1) $X \in C^1[t_0, \infty)$ (as a fundamental matrix),
2) $\|X(t)\| \leqslant C$ (by the first assumption of the theorem),
3) $\|\dot{X}(t)\| \leqslant \|A(t)\|\|X(t)\| \leqslant MC$,
4) the boundedness of the inverse matrix follows from the fact that $\operatorname{Det} X(t)$ is bounded away from zero.

Indeed, by (3.2.2) we have

$$\begin{aligned} |\operatorname{Det} X(t)| &= |\operatorname{Det} X(t_0)| \left| \exp \int_{t_0}^{t} \operatorname{Sp} A(\tau)\, d\tau \right| \\ &\geqslant |\operatorname{Det} X(t_0)| e^{a} > 0. \end{aligned}$$

The transformation $x = X(t)y$ reduces system (3.1.1) to the system $\dot{y} = B(t)y$. We determine $B(t)$ according to formula (3.1.4):

$$\begin{aligned} B(t) &= X^{-1}(t)A(t)X(t) - X^{-1}(t)\dot{X}(t) \\ &= X^{-1}(t)A(t)X(t) - X^{-1}(t)A(t)X(t) \equiv 0. \qquad \square \end{aligned}$$

COROLLARY 3.2.1. *If the matrix* $A(t)$ *is absolutely integrable, i.e.,*

$$\text{(3.2.3)} \qquad \int_{t_0}^{\infty} \|A(\tau)\|\, d\tau = k \leqslant \infty,$$

*then system* (3.1.1) *is reducible to the system with zero matrix.*

PROOF. Let us show that both conditions of the previous theorem are satisfied.

1. Any solution $x(t)$ satisfies the integral equation

$$x(t) = x(t_0) + \int_{t_0}^{t} A(\tau)x(\tau)\, d\tau,$$

or

$$\|x(t)\| \leqslant \|x(t_0)\| + \int_{t_0}^{t} \|A(\tau)\|\|x(\tau)\|\, d\tau.$$

Hence, by the Gronwall-Bellman lemma (see Appendix) we have

$$\text{(3.2.4)} \qquad \|x(t)\| \leqslant \|x(t_0)\| e^{\int_{t_0}^{t} \|A(\tau)\|\, d\tau} \leqslant \|x(t_0)\| e^{k}.$$

Inequality (3.2.4) guarantees that the first condition of the theorem is satisfied.

2. We have

$$\left| \int_{t_0}^{t} \operatorname{Sp} A(\tau)\, d\tau \right| \leqslant \int_{t_0}^{t} |\operatorname{Sp} A(\tau)|\, d\tau \leqslant n \int_{t_0}^{t} \|A(\tau)\|\, d\tau \overset{(3.2.3)}{\leqslant} nk.$$

Thus,

$$\int_{t_0}^{t} \operatorname{Re} \operatorname{Sp} A(\tau)\, d\tau \geqslant -nk > -\infty. \quad \square$$

COROLLARY 3.2.2. *For the characteristic exponent of any solution $x(t)$ of system* (3.1.1) *the estimate*

$$\chi[x] \leqslant \overline{\lim_{t\to\infty}} \frac{1}{t} \int_{t_0}^{t} \|A(\tau)\|\, d\tau \tag{3.2.5}$$

*is valid* (*this follows from* (3.2.4)).

## §3. Reduction of a linear system to a triangular or a block-triangular form

At the beginning of this section we shall deal with the Perron transformation. This is a unitary transformation reducing a system

$$\dot{x} = A(t)x, \qquad x \in \mathbb{C}^n, \quad A \in C[t_0, \infty), \tag{3.3.1}$$

to an upper-triangular system with real diagonal coefficients. Note that here the boundedness of $\|A(t)\|$ is not required.

LEMMA 3.3.1. *Any fundamental matrix $X(t)$ of system* (3.3.1) *can be represented in the form of the product of two continuously differentiable matrices: a unitary one $U(t)$ and an upper triangular one $R(t)$ with positive diagonal.*

REMARK 3.1.1. This result is proved and widely used in linear algebra. Let us reproduce the proof since it provides an explicit form of $R(t)$ in terms of $X(t)$; this is used in what follows.

PROOF. Let

$$X(t) = \{x_1(t), \dots, x_n(t)\}.$$

We apply the Schmidt orthogonalization process to the basis vectors:

$$\begin{aligned}
&\xi_1 = x_1, && e_1 = \xi_1/\|\xi_1\|,\\
&\xi_2 = x_2 - (x_2, e_1)e_1, && e_2 = \xi_2/\|\xi_2\|,\\
&\dots\dots\dots\dots && \dots\dots\dots\dots\\
&\xi_n = x_n - \sum_{s=1}^{n-1}(x_n, e_s)e_s, && e_n = \xi_n/\|\xi_n\|.
\end{aligned}$$

Obviously, $(e_i, e_j) = \delta_{ij}$, i.e., the matrix

$$U(t) = \{e_1, \dots, e_n\}$$

is unitary since $U^*U = E$. At the same time

$$\begin{aligned}
x_1 &= \|\xi_1\|e_1,\\
x_2 &= (x_2, e_1)e_1 + \|\xi_2\|e_2,\\
&\dots\dots\dots\dots\\
x_n &= \sum_{s=1}^{n-1}(x_n, e_s)e_s + \|\xi_n\|e_n.
\end{aligned}$$

This implies that the equality

$$X(t) = U(t)R(t) \tag{3.3.2}$$

holds, where

$$R(t) = \begin{pmatrix} \|\xi_1\| & (x_2, e_1) & \dots & (x_n, e_1) \\ 0 & \|\xi_2\| & \dots & (x_n, e_2) \\ \vdots & & & \vdots \\ 0 & \dots & \dots & \|\xi_n\| \end{pmatrix}, \tag{3.3.3}$$

i.e.,

$$r_{ij}(t) = 0, \qquad i > j, \qquad r_{ii}(t) > 0 \qquad \text{for } t \geqslant t_0, \quad i, j = 1, \dots, n. \quad \square$$

THEOREM 3.3.1 (Perron's theorem on the triangulation of a linear system). *By means of a unitary transformation any linear system* (3.3.1) *can be reduced to a system with an upper triangular matrix whose diagonal coefficients are real. If the initial system has bounded coefficients, then the coefficients of the triangular system are also bounded, and the unitary transformation is Lyapunov.*

PROOF. We show that there exists a transformation

$$x = U(t)y, \qquad U^*(t)U(t) = E, \tag{3.3.4}$$

such that

$$\dot{y} = (U^{-1}(t)A(t)U(t) - U^{-1}(t)\dot{U}(t))y \equiv B(t)y, \tag{3.3.5}$$

where

$$b_{kj}(t) = 0, \qquad k > j, \qquad \operatorname{Im} b_{kk}(t) = 0.$$

Let us take a fundamental matrix $X(t)$ of system (3.3.1) and choose the unitary matrix defined by Lemma 3.3.1 as $U(t)$. The transformation (3.3.4) for the matrix $X(t)$ uniquely defines a fundamental matrix $Y(t)$ of system (3.3.5),

$$X(t) = U(t)Y(t).$$

From condition (3.3.2) we obtain that $Y(t) \equiv R(t)$ is upper triangular with positive diagonal. From system (3.3.5) we have $\dot{Y}(t) \equiv B(t)Y(t)$, or

$$B(t) = \dot{Y}(t)Y^{-1}(t) = \dot{R}(t)R^{-1}(t),$$

i.e., $B(t)$ is upper triangular and

$$b_{kk}(t) = \dot{r}_{kk}(t)r_{kk}^{-1}(t) = \frac{\|\dot{\xi}_k(t)\|}{\|\xi_k(t)\|} = \frac{d}{dt}\ln\|\xi_k(t)\|,$$

which follows from (3.3.3). Thus the first part of the theorem is proved. Now we show that $\|B(t)\|$ is bounded for $t \geqslant t_0$ if $\|A(t)\|$ is bounded. Indeed,

$$B(t) = U^{-1}AU - U^{-1}\dot{U} = \widetilde{A}(t) - V(t).$$

If $\sup_{t \geqslant t_0} \|A(t)\| \leqslant M$, then

$$\|\widetilde{A}(t)\| \leqslant \|U^{-1}(t)\| \cdot \|A(t)\| \cdot \|U(t)\| \leqslant M \qquad \text{for} \qquad t \geqslant t_0.$$

Note that $V^*(t) = -V(t)$, i.e., $V(t)$ is a skew-Hermitian matrix. Indeed, let us verify the equality

$$(U^{-1}\dot{U})^* = -U^{-1}\dot{U}. \tag{3.3.6}$$

By differentiating the identity $U^*U = E$, we obtain

$$\dot{U}^*U + U^*\dot{U} = 0;$$

therefore,

$$\dot{U} = -(U^*)^{-1}\dot{U}^*U.$$

Substituting $\dot{U}$ in the left-hand side of equality (3.3.6) and taking into account that $U^*U = E$, we find that (3.3.6) holds. The diagonal elements of skew-Hermitian matrices are purely imaginary; therefore,

$$V_{kk}(t) = \operatorname{Im}\widetilde{a}_{kk}(t)$$

since

$$\operatorname{Im} b_{kk}(t) = 0, \qquad k = 1, \dots, n, \quad t \geqslant t_0.$$

$B(t)$ is upper triangular; thus,

$$V_{kj}(t) = \widetilde{a}_{kj}(t) \qquad \text{for} \qquad k > j,$$

and, since $V$ is skew-Hermitian, we have

$$V_{kj}(t) = -\overline{V_{jk}(t)} \qquad \text{for} \qquad k < j.$$

Hence, $\|V(t)\| \leqslant M$, and, therefore,

$$\|B(t)\| \leqslant \|\widetilde{A}(t)\| + \|V(t)\| \leqslant 2M, \qquad t \geqslant t_0.$$

Let us show that the matrix $U(t)$ of a Perron transformation is Lyapunov if the coefficients of the system are bounded. Indeed, $\|U(t)\|$ and $\|U^{-1}(t)\|$ are bounded since $U$ is unitary. The boundedness of $\|\dot{U}(t)\|$ follows from the boundedness of $V(t)$, since $\dot{U}(t) = U(t)V(t)$. □

REMARK 3.3.2. If in system (3.3.1) $A(t)$ is real, then $U(t)$ can be chosen orthogonal.

The unitary matrix $U(t)$ realizing the triangulation of the linear system (3.3.1) is constructed, as we have seen in the Perron theorem, in accordance with the chosen fundamental matrix $X(t)$. A triangular system can be successively integrated and, naturally, one is led to an optimistic conjecture that there exists a unitary transformation that is not connected with fundamental matrices. As the following theorem shows, there is no reason for optimism.

THEOREM 3.3.2 (on the inversion of Perron's theorem). *Any unitary transformation reducing the linear system (3.3.1) to a triangular one with real diagonal is a Perron transformation, i.e., it can be obtained from some fundamental matrix by the Schmidt orthogonalization process.*

PROOF. Let some unitary transformation (3.3.4) reduce system (3.3.1) to a triangular system (3.3.5) with the properties of the Perron theorem. Consider a fundamental matrix $Y(t, t_0)$ of this system. This is an upper triangular matrix with positive diagonal, since

$$y_{kk}(t, t_0) = \exp \int_{t_0}^{t} b_{kk}(\tau)\, d\tau, \qquad k = 1, \dots, n.$$

Note that the matrix $Y^{-1}(t, t_0)$ also has these properties. We form a fundamental matrix $X(t)$ of system (3.3.1), $X(t) = U(t)Y(t, t_0)$; therefore,

$$U(t) = X(t)Y^{-1}(t, t_0).$$

It is known from algebra **[16]** that for any nonsingular matrix $X(t)$ there exists a unique triangular matrix $S(t)$ with positive diagonal such that $X(t)S(t)$ is unitary. Comparing this result with (3.3.2), we see that $U(t)$ is a Perron matrix. □

In some cases, system (3.3.1) can be reduced to a block-triangular form by means of a Lyapunov transformation, i.e., the system can be divided into $r$ independent triangular systems. Let us find conditions for this to be possible.

THEOREM 3.3.3. *If a linear system* (3.3.1) *has a fundamental matrix*

$$X(t) = \{X_{n_1}(t), \dots, X_{n_r}(t)\}, \qquad n_1 + n_2 + \cdots + n_r = n, \tag{3.3.7}$$

*such that*

$$\inf_t \frac{G(X)}{G(X_{n_1})G(X_{n_2})\dots G(X_{n_r})} = \rho > 0, \tag{3.3.8}$$

*then there exists a Lyapunov transformation reducing the system to block-triangular form with real diagonal*:

$$\dot{y} = \operatorname{diag}[B_{n_1}(t), \dots, B_{n_r}(t)]y \equiv B(t)y. \tag{3.3.9}$$

REMARK 3.3.3. The notation (3.3.7) indicates that the set of basis solutions

$$x_1(t), x_2(t), \dots, x_n(t)$$

constituting the fundamental matrix $X(t)$ is divided into $r$ blocks, and the block $X_{n_i}(t)$ contains $n_i$ solutions. $G(X_{n_i})$ is the Gram determinant of the set of vector-solutions entering the block $X_{n_1}$. Each matrix $B_{n_i}(t)$ is upper triangular, $i = 1, \dots, r$.

PROOF. For every set $X_{n_i}(t)$ of vectors in the fundamental matrix $X(t)$, $i = 1, \dots, r$, we carry out the Schmidt orthogonalization process separately, i.e., having exhausted one set, we start the process anew for the next set. Thus, we obtain the representation

$$X(t) = (U_{n_1}(t), \dots, U_{n_r}(t)) \operatorname{diag}[R_{n_1}(t), \dots, R_{n_r}(t)], \tag{3.3.10}$$

where $U^*_{n_i}(t)U_{n_i}(t) \equiv E_{n_i}$, and $R_{n_i}(t)$ is an $n_i \times n_i$ upper triangular matrix with positive diagonal. This result immediately follows from the proof of the Perron theorem. Let us consider the transformation

$$x = (U_{n_1}(t), \dots, U_{n_r}(t))y \equiv V(t)y \tag{3.3.11}$$

and show that it is the required one. Indeed, for the fundamental matrices $X(t)$ of system (3.3.1) and $Y(t)$ of the system

$$\dot{y} = (V^{-1}AV - V^{-1}\dot{V})y \equiv B(t)y$$

we have $X(t) = V(t)Y(t)$, or, according to (3.3.10),

$$Y(t) \equiv \mathrm{diag}[R_{n_1}(t), \dots, R_{n_r}(t)].$$

Therefore,

$$B(t) = \dot{Y}(t)Y^{-1}(t),$$

i.e., the matrix $B(t)$ has the form indicated in formula (3.3.9). Now let us show that the matrix $V(t)$ is Lyapunov. By construction, the norm of each of its columns is equal to unity, i.e., $\|V(t)\|$ is bounded. The boundedness of $\|\dot{V}(t)\|$ follows from the fact that any block $V_{n_i}(t)$ consists of the first $n_i$ columns of a Perron matrix, and the derivative of the latter is bounded. It remains to show that $\|V^{-1}(t)\|$ is bounded, or that $\mathrm{Det}\, V(t)$ is bounded away from zero. We write $V(t)$ in the following way (see (3.3.11)):

$$V(t) = X(t)S(t) \equiv X(t)\, \mathrm{diag}\left[R_{n_1}^{-1}(t), \dots, R_{n_r}^{-1}(t)\right].$$

Then

$$\begin{aligned} |\,\mathrm{Det}\, V(t)|^2 &= \mathrm{Det}(V^*V) = \mathrm{Det}(S^*X^*XS) \\ &= \mathrm{Det}(S^*S)\,\mathrm{Det}(X^*X) \overset{(3.3.10)}{\geqslant} \rho\, \mathrm{Det}(S^*S) G(X_{n_1}) \dots G(X_{n_r}) \\ &= \rho \prod_{i=1}^{r} \mathrm{Det}(S_{n_i}^* S_{n_i})\, \mathrm{Det}(X_{n_i}^* X_{n_i}) \\ &= \rho \prod_{i=1}^{r} \mathrm{Det}(S_{n_i}^* X_{n_i}^* X_{n_i} S_{n_i}) \\ &= \rho \prod_{i=1}^{r} \mathrm{Det}(U_{n_i}^* U_{n_i}) = \rho. \quad \square \end{aligned}$$

The converse theorem also holds.

THEOREM 3.3.4. *If system* (3.3.1) *is reducible by a Lyapunov transformation to a system of the form* (3.3.9), *then it has a fundamental matrix* $X(t)$ *that satisfies condition* (3.3.8).

PROOF. Let a Lyapunov transformation $x = L(t)y$ transform (3.3.1) into (3.3.9). The latter system can be successively integrated and its fundamental matrix

$$Y(t, t_0) = \mathrm{diag}[Y_{n_1}(t), \dots, Y_{n_r}(t)]$$

is block-triangular. Define

$$X_{n_i}(t) = L(t)Y_{n_i}(t), \qquad i = 1, \dots, r.$$

Thus,

$$X(t) = \{X_{n_1}(t), \dots, X_{n_r}(t)\} = \{L_{n_1}(t), \dots, L_{n_r}(t)\}\, \mathrm{diag}[Y_{n_1}(t), \dots, Y_{n_r}(t)].$$

Hence,

$$\begin{aligned} G(X_{n_i}) &= \mathrm{Det}(X_{n_i}^* X_{n_i}) \\ &= \mathrm{Det}\{ Y_{n_i}^* L_{n_i}^* L_{n_i} Y_{n_i} \} \\ &= \exp\left(2 \int_{t_0}^{t} \mathrm{Sp}\, B_{n_i}(\tau)\, d\tau\right) G(L_{n_i}), \\ G(X) &= |\,\mathrm{Det}\, X|^2 = |\,\mathrm{Det}\, L|^2 |\,\mathrm{Det}\, Y|^2 \\ &= e^{2\int_{t_0}^{t} \mathrm{Sp}\, B(\tau)\, d\tau} G(L), \end{aligned}$$

or

$$\begin{aligned} \frac{G(X)}{G(X_{n_1}) \dots G(X_{n_r})} &= \frac{\exp\left(2 \int_{t_0}^{t} \mathrm{Sp}\, B(\tau)\, d\tau\right) G(L)}{G(L_{n_1}) \dots G(L_{n_r}) \exp\left(2 \int_{t_0}^{t} \sum_{i=1}^{r} \mathrm{Sp}\, B_{n_i}\, d\tau\right)} \\ &= \frac{G(L)}{G(L_{n_1}) \dots G(L_{n_r})}. \end{aligned}$$

By the properties of Lyapunov matrices, there exist constants $a$ and $b$ such that

$$G(L) \geqslant a, \qquad G(L_{n_i}(t)) \leqslant b, \qquad i = 1, \dots, r, \qquad \text{for} \qquad t \geqslant t_0;$$

consequently,

$$\frac{G(X)}{G(X_{n_1}) \dots G(X_{n_r})} \geqslant \frac{a}{b^r},$$

i.e., the condition (3.3.8) is satisfied. □

COROLLARY 3.3.1. *The condition* (3.3.8) *is necessary and sufficient for the system* (3.3.1) *to be reducible to the block-triangular form* (3.3.9).

COROLLARY 3.3.2. *For the system* (3.3.1) *to be reducible to diagonal form it is necessary and sufficient that it have a fundamental matrix* $X(t)$ *such that*

$$\frac{G(X)}{\|x_1(t)\|^2 \dots \|x_n(t)\|^2} \geqslant \rho > 0, \qquad t \geqslant t_0. \tag{3.3.12}$$

EXAMPLE 3.3.1. a) The system

$$\begin{aligned} \dot{x} &= 5x - y - 4z, \\ \dot{y} &= -12x + 5y + 12z, \\ \dot{z} &= 10x - 3y - 9z \end{aligned}$$

is reducible to block-triangular form consisting of two blocks of dimensions 1 and 2, respectively. Indeed,

$$X(t) = \begin{pmatrix} e^{-t} & e^t & (t+1)e^t \\ -2e^{-t} & 0 & 3e^t \\ 2e^{-t} & e^t & te^t \end{pmatrix} \equiv \{X_{n_1}(t), X_{n_2}(t)\}, \qquad n_1 = 1, \quad n_2 = 2;$$

$$\frac{G(x)}{G(X_{n_1}) \cdot G(X_{n_2})} = \frac{e^{2t}}{9e^{-2t} \cdot 19e^{4t}} = \frac{1}{9 \cdot 19}.$$

The condition (3.3.8) is satisfied.

EXAMPLE 3.3.2.

$$\begin{aligned} \dot{x}_1 &= -x_1, \\ \dot{x}_2 &= (\sin \ln t + \cos \ln t)x_2, \end{aligned} \qquad t \geqslant 1.$$

We have

$$X(t) = \begin{pmatrix} e^{-t} & 0 \\ 0 & e^{t \sin \ln t} \end{pmatrix} = \{x_1(t), x_2(t)\},$$

$$\chi[x_1] = -1, \qquad \chi[x_2] = 1.$$

Let us verify the condition (3.3.17) for $x_2(t)$:

$$e^{t \sin \ln t - s \sin \ln s} \overset{?}{\leqslant} a e^{\lambda(t-s)}, \qquad \lambda > 0, \quad a > 0, \quad t \leqslant s.$$

Choose the sequences

$$s_k = \exp\left(2k + \frac{3}{2}\right)\pi,$$

$$t_k = \exp\left(2k + \frac{1}{2}\right)\pi.$$

Thus,

$$t_k - s_k = (1 - e^{\pi}) \exp\left(2k + \frac{1}{2}\right)\pi \underset{k\to\infty}{\longrightarrow} -\infty.$$

At the same time,

$$t_k \sin \ln t_k - s_k \sin \ln s_k = (1 + e^{\pi}) \exp\left(2k + \frac{1}{2}\right)\pi \underset{k\to\infty}{\longrightarrow} \infty.$$

Therefore, the condition (3.3.17) is not satisfied for $x_2(t)$.

THEOREM 3.3.6. *A hyperbolic system is reducible to block-triangular form consisting of two blocks whose diagonals are real.*

PROOF. If we show that

$$\alpha(t) = \sphericalangle(L^+(t), L^-(t)) \geqslant \rho > 0,$$

then inequality (3.3.15) will be satisfied for the basis sets $X$ and $Y$ of the subspaces $L^+$ and $L^-$; thus, the condition of Theorem 3.3.3, which implies the required result, will also be satisfied.

We prove this by contradiction. Let, conversely, there exist a sequence

$$t_i \to \infty, \qquad i \to \infty,$$

such that

$$\alpha(t_i) \to 0, \qquad i \to \infty.$$

At each moment $t_i$ the angle $\alpha(t_i)$ is realized by a pair of vectors $x_i(t_i) \in L^+(t_i)$ and $y_i(t_i) \in L^-(t_i)$, and, without loss of generality, we can assume that

$$\|x_i(t_i)\| = 1, \qquad \|y_i(t_i)\| = 1.$$

It can be easily seen that

$$\|x_i(t_i) - y_i(t_i)\| \to 0, \qquad i \to \infty.$$

According to the theorem on integral continuity [**5**],

$$\|y_i(t_i + T) - x_i(t_i + T)\| \underset{t\to\infty}{\longrightarrow} 0 \tag{3.3.18}$$

holds for any fixed $T > 0$. At the same time,

$$y_i(t) \in L^-(t), \qquad x_i(t) \in L^+(t),$$

and, by the estimates (3.3.17) and (3.3.18), we have

$$1 = \|y_i(t_i, t_i + T, y_i(t_i + T))\| \leqslant a\|y_i(t_i + T)\|e^{-\lambda T},$$

$$\|x_i(t_i + T, t_i, x(t_i))\| \leqslant ae^{-\lambda T}\|x_i(t_i)\| = ae^{-\lambda T},$$

or

$$\|y_i(t_i + T)\| \geqslant e^{\lambda T}/a, \qquad \|x_i(t_i + T)\| \leqslant ae^{-\lambda T},$$

and, finally, we have

$$\begin{aligned}\|y_i(t_i + T) - x_i(t_i + T)\| &\geqslant \|y_i(t_i + T)\| - \|x_i(t_i + T)\| \\ &\geqslant e^{\lambda T}/a - ae^{-\lambda T} \\ &= e^{\lambda T}(1 - a^2e^{-2\lambda T})/a.\end{aligned}$$

The last quantity does not depend on $i$ and by choosing $T > 0$ it can be made, e.g., greater than one; this contradicts (3.3.18). Thus, we have come to a contradiction with the assumption that $\inf_t \alpha(t) = 0$. □

## §4. Almost reducible systems

DEFINITION 3.4.1. A system

$$x = A(t)x \tag{3.4.1}$$

is said to be *almost reducible* to a system

$$\dot{y} = B(t)y, \tag{3.4.2}$$

if for any $\delta > 0$ there exists a Lyapunov transformation

$$x = L_\delta(t)y \tag{3.4.3}$$

reducing system (3.4.1) to a system

$$\dot{y} = [B(t) + \Phi(t)]y, \tag{3.4.4}$$

where

$$\|\Phi(t)\| \leqslant \delta \qquad \text{for} \qquad t \geqslant t_0. \tag{3.4.5}$$

Recall that

$$L_\delta^{-1}(t)A(t)L_\delta(t) - L_\delta^{-1}(t)\dot{L}_\delta(t) = B(t) + \Phi(t). \tag{3.4.6}$$

The notion of almost reducibility was introduced by Bylov **[10]**.

Note that for any fixed $\delta > 0$ the functions

$$\|L_\delta(t)\|, \qquad \|\dot{L}_\delta(t)\|, \qquad \|L_\delta^{-1}(t)\|$$

are bounded for $t \geqslant t_0$ since $L_\delta(t)$ is a Lyapunov matrix. However, they can grow unboundedly as $\delta \to 0$. This means that almost reducibility does not imply reducibility. The converse is obviously true.

Note the following properties of almost reducibility.

1. *Transitivity*. If a system $\dot{x} = A(t)x$ is almost reducible to a system $\dot{y} = B(t)y$ and $\dot{y} = B(t)y$ is almost reducible to $\dot{z} = C(t)z$, then the first system is almost reducible to the last one.

PROOF. For $\alpha > 0$, $\beta > 0$ from the conditions (3.4.3)–(3.4.6) we have

(3.4.7)
$$L_\alpha^{-1}(t)A(t)L_\alpha(t) - L_\alpha^{-1}\dot{L}_\alpha(t) = B(t) + \Phi_1(t), \qquad \|\Phi_1(t)\| \leqslant \alpha,$$

(3.4.8)
$$L_\beta^{-1}(t)B(t)L_\beta(t) - L_\beta^{-1}(t)\dot{L}_\beta(t) = C(t) + \Phi_2(t), \qquad \|\Phi_2(t)\| \leqslant \beta.$$

Let us show that for any $\delta > 0$ we can choose $\alpha > 0$, $\beta > 0$ such that the transformation $x = L_\alpha(t)L_\beta(t)z$ almost reduces the system $\dot{x} = A(t)x$ to the system $\dot{z} = C(t)z$.

Consider the coefficient matrix of the new system, taking (3.4.6) into account:

$$\begin{aligned} L_\beta^{-1}(t)L_\alpha^{-1}(t)A(t)L_\alpha(t)L_\beta(t) &- L_\beta^{-1}(t)L_\alpha^{-1}(t)\left(\dot{L}_\alpha(t)L_\beta(t) + L_\alpha(t)\dot{L}_\beta(t)\right) \\ &= L_\beta^{-1}(t)\left(L_\alpha^{-1}(t)A(t)L_\alpha(t) - L_\alpha^{-1}(t)\dot{L}_\alpha(t)\right)L_\beta(t) - L_\beta^{-1}(t)\dot{L}_\beta(t) \\ &\overset{(3.4.7)}{=} L_\beta^{-1}(t)\left(B(t) + \Phi_1(t)\right)L_\beta(t) - L_\beta^{-1}(t)\dot{L}_\beta(t) \\ &\overset{(3.4.8)}{=} C(t) + L_\beta^{-1}(t)\Phi_1(t)L_\beta(t) + \Phi_2(t). \end{aligned}$$

Now let us choose $\delta > 0$ and set

$$\beta = \delta/2, \qquad \alpha = \delta/(2M^2),$$

where

$$M = \sup_{t \geqslant t_0}\{\|L_\beta^{-1}(t)\|, \|L_\beta(t)\|\}.$$

Therefore,

$$\|L_\beta^{-1}(t)\Phi_1(t)L_\beta(t) + \Phi_2(t)\| \leqslant M^2\delta/(2M^2) + \delta/2 = \delta. \quad \square$$

2. Almost reducibility (in contrast to reducibility) does not possess symmetry.

Millionshchikov indicated **[30]** the existence of two linear systems, one of which is amost reducible to the other but not vice versa. These arguments required complicated and subtle methods of the theory of linear systems, not yet touched upon here.

3. If system (3.4.1) is almost reducible to (3.4.2), then the system $\dot{x} = -A^*(t)x$ is almost reducible to the system $\dot{y} = -B^*(t)y$.

The statement is verified straightforwardly (see §2).

We pass to the exposition of concrete results of the theory of almost reducibility and standard transformations used there.

THEOREM 3.4.1. *Any linear system is almost reducible to some diagonal system with real coefficients.*

PROOF. Let there be given a system $\dot{x} = A(t)x$ with a continuous and bounded for $t \geqslant t_0$ matrix $A(t)$. In this case the Perron transformation (Theorem 3.3.1) is Lyapunov and reduces our system to a system $\dot{y} = B(t)y$, where

$$b_{jk}(t) = 0 \qquad \text{for} \qquad j > k, \qquad \operatorname{Im} b_{ii}(t) = 0, \qquad i = 1, \dots, n.$$

Thus, a real diagonal is obtained. Now let us turn to a standard, so-called *β-transformation*, which allows one to make the off-diagonal elements of a triangular matrix arbitrarily small. Let $y = Sz$, where

$$S = \operatorname{diag}[1, \beta, \beta^2, \dots, \beta^{n-1}].$$

Obviously, $S$ is a Lyapunov matrix and the system $\dot{z} = C(t)z$ is such that

$$C(t) = S^{-1}B(t)S = \begin{pmatrix} b_{11} & \beta b_{12} & \dots & \beta^{n-1}b_{1n} \\ 0 & b_{22} & & \vdots \\ \vdots & & \ddots & \beta b_{n-1,n} \\ 0 & \dots & \dots & b_{nn} \end{pmatrix}$$

$$= \operatorname{diag}[b_{11}(t), \dots, b_{nn}(t)] + \Phi(t).$$

Since $\sup_{t \geqslant t_0} \|B(t)\| < \infty$, we can choose $\beta > 0$ for any $\delta > 0$ so small that the inequality $\|\Phi(t)\| \leqslant \delta$ is satisfied for $t \geqslant t_0$.

It remains to use the transitivity of almost reducibility. □

REMARK 3.4.1. In Theorem 3.4.1 we described the $\beta$-transformation and applied it to a triangular matrix. If we consider a square matrix $B(t)$, then

$$S^{-1}B(t)S = \begin{pmatrix} b_{11} & \beta b_{12} & \dots & \beta^{n-1}b_{1n} \\ b_{21}/\beta & b_{22} & \dots & \beta^{-2}b_{2n} \\ \dots & \dots & \dots & \dots \\ b_{n1}/\beta^{n-1} & \dots & b_{n,n-1}/\beta & b_{nn} \end{pmatrix}.$$

In this case the decrease of one corner of the matrix at the expense of the choice of $\beta$ results in the increase of the other and vice versa.

THEOREM 3.4.2. *If a linear system is almost reducible to some diagonal system $\dot{u} = P(t)u$, then it is also almost reducible to the system $\dot{z} = \operatorname{Re} P(t)z$.*

PROOF. Let

$$P(t) = R(t) + iQ(t),$$

where $R(t)$ and $Q(t)$ are now real matrices. Let us consider the matrix

$$L(t) = \exp\left(i \int_{t_0}^{t} Q(\tau)\, d\tau\right)$$

and show that this matrix is Lyapunov. Indeed,

$$L^*(t) = \exp\left(-i \int_{t_0}^{t} Q(t)\, d\tau\right) = L^{-1}(t),$$

i.e., the matrix is unitary; therefore, it is bounded together with its inverse. Moreover,

$$\dot{L}(t) = iL(t)Q(t)$$

is bounded under the condition that $Q(t)$ is bounded. The transformation $u = L(t)z$ leads to the system with the coefficient matrix

$$L^{-1}(t)P(t)L(t) - L^{-1}(t)\dot{L}(t) = P(t) - iQ(t) = R(t).$$

It remains to use the transitivity. □

REMARK 3.4.2. The transformation described in the theorem is called Re-*transformation*. We have applied it to a diagonal matrix $P(t)$. Consider the general case. Let there be given a system $\dot{x} = A(t)x$, where $A_d = R(t) + iQ(t)$. The transformation

$$x = \exp\left(i \int_{t_0}^{t} Q(\tau)\, d\tau\right) y$$

is Lyapunov if $A(t)$ is bounded (as shown above) and

$$(L^{-1}(t)A(t)L(t) - L^{-1}(t))_d = \operatorname{Re} A_d.$$

Thus, this transformation annihilates the imaginary part of the diagonal of the matrix of coefficients.

THEOREM 3.4.3. *If a linear system is almost reducible to a system*

$$\dot{y} = By$$

*with a constant matrix $B$, then it is almost reducible to the system*

$$\dot{y} = \operatorname{diag}[\operatorname{Re}\lambda_1, \dots, \operatorname{Re}\lambda_n]y,$$

*where $\lambda_1, \dots, \lambda_n$ are the eigenvalues of the matrix $B$.*

PROOF. The proof is left to the reader. One has to trace subsequent transformations from the initial system to the final one and then use the transitivity. □

It follows from the above-given theorems that if system (3.4.1) is almost reducible to (3.4.2), then we can assume that the matrix $B(t)$ is real diagonal without loss of generality.

THEOREM 3.4.4. *If system (3.4.1) is almost reducible to system (3.4.2) with a diagonal matrix $B(t)$, then it is also almost reducible to a system with matrix $\widetilde{B}(t)$, obtained from $B(t)$ by means of any rearrangement of its elements.*

PROOF. The diagonal matrices $B(t)$ and $\widetilde{B}(t)$ are similar and the similarity transformation is carried out by a constant matrix, which is Lyapunov. □

## §5. Regular systems

Consider a system

$$\dot{x} = A(t)x, \tag{3.5.1}$$

where

$$x \in \mathbb{C}^n, \qquad A \in C[t_0, \infty), \qquad \sup_{t \geqslant 0} \|A(t)\| \leqslant M.$$

We introduce the notation

$$\Lambda = \chi\left[\exp \int_{t_0}^{t} \operatorname{Sp} A(\tau)\, d\tau\right], \qquad \Lambda' = \chi\left[\exp\left(-\int_{t_0}^{t} \operatorname{Sp} A(\tau)\, d\tau\right)\right]. \tag{3.5.2}$$

According to Theorem 2.1.2 on the characteristic exponent of a product, we have $\Lambda + \Lambda' \geqslant 0$. Using the Lyapunov inequality (2.5.1), we write

$$\sigma_X \geqslant \Lambda \geqslant -\Lambda'. \tag{3.5.3}$$

DEFINITION 3.5.1. A linear homogeneous system (3.5.1) is said to be *regular* if the equality $\sigma_{\bar{X}} = -\Lambda'$ holds for some fundamental matrix $\bar{X}(t)$ of this system. In the opposite case this system is said to be *irregular*.

Note that the fundamental matrix $\bar{X}(t)$ is necessarily normal, since the Lyapunov equality is satisfied for it. If $\{\alpha_1, \alpha_2, \dots, \alpha_n\}$ is the spectrum of system (3.5.1), then the regularity of this system is equivalent to the equality

$$\sum_{i=1}^{n} \alpha_i = -\Lambda'.$$

The definition of a regular system is due to Lyapunov **[25]**. Let us clarify its meaning.

LEMMA 3.5.1. *A linear homogeneous system* (3.5.1) *is regular if and only if the following two conditions are satisfied simultaneously*:

1) *the limit*

$$\lim_{t\to\infty} \frac{1}{t} \int_{t_0}^{t} \operatorname{Re} \operatorname{Sp} A(\tau)\, d\tau = S \tag{3.5.4}$$

*exists;*

2) *there exists a fundamental matrix* $X(t)$ *such that* $\sigma_X = S$.

PROOF. *Necessity.* The regularity of (3.5.1) implies that for some basis $X(t)$ the equality $\sigma_X = -\Lambda'$ is satisfied; thus, by (3.5.3) we have $\Lambda = -\Lambda'$, i.e., (3.5.4) holds; and the second condition of the lemma also holds since $-\Lambda' = S$.

*Sufficiency.* It follows from the first condition of the lemma that $\Lambda = -\Lambda' = S$ and from the second that $\sigma_X = -\Lambda'$. □

EXAMPLE 3.5.1. We borrow an example of an irregular system from **[25]**:

$$\dot{x}_1 = x_1 \cos \ln t + x_2 \sin \ln t,$$
$$\dot{x}_2 = x_1 \sin \ln t + x_2 \cos \ln t.$$

The system

$$X(t) = \{x_1(t), x_2(t)\} = \begin{pmatrix} e^{t \sin \ln t} & e^{t \cos \ln t} \\ e^{t \sin \ln t} & -e^{t \cos \ln t} \end{pmatrix}$$

is a normal fundamental system;

$$\sigma_X = \chi[x_1] + \chi[x_2] = 1 + 1 = 2.$$

At the same time

$$\Lambda' = \chi\left[e^{-\int_{t_0}^{t} 2 \cos \ln \tau \, d\tau}\right] = \chi\left[e^{t(\sin \ln t + \cos \ln t)}\right] = \sqrt{2}.$$

Hence, $\sigma_X \neq -\Lambda'$, i.e., the system is irregular.

LEMMA 3.5.2. *A Lyapunov transformation preserves the regularity of a system.*

PROOF. Let a regular system (3.5.1) be reduced to the system

$$\dot{y} = [L^{-1}(t)A(t)L(t) - L^{-1}(t)\dot{L}(t)]y \equiv B(t)y \tag{3.5.5}$$

by means of the Lyapunov transformation $x = L(t)y$. Lyapunov transformations do not change the spectrum $\{\alpha_1, \alpha_2, \dots, \alpha_n\}$ of the system; therefore, the regularity of (3.5.5) follows from the equality

$$\sum_{i=1}^{n} \alpha_i = \chi \left[ \exp\left( - \int_{t_0}^{t} \operatorname{Sp} B(\tau) \right) \right]. \tag{3.5.6}$$

Consider the right-hand side of (3.5.6). According to Theorem 2.1.4, we have

$$\chi \left[ e^{-\int_{t_0}^{t} \operatorname{Sp}[L^{-1}AL - L^{-1}\dot{L}]\, d\tau} \right] = \Lambda' + \chi \left[ e^{\int_{t_0}^{t} \operatorname{Sp} L^{-1}\dot{L}\, d\tau} \right].$$

The second term on the right-hand side of the equality obtained is equal to zero as the characteristic exponent of a bounded function. Indeed, for a Lyapunov matrix we have $\dot{L} \equiv \dot{L}L^{-1}L$; therefore,

$$\operatorname{Det} L(t)/\operatorname{Det} L(t_0) = e^{\int_{t_0}^{t} \operatorname{Sp}[\dot{L}(\tau)L^{-1}]\, d\tau} = e^{\int_{t_0}^{t} \operatorname{Sp}[L^{-1}(\tau)\dot{L}(\tau)\, d\tau}.$$

Hence, (3.5.6) is valid, since it is nothing else but

$$\sum_{i=1}^{n} \alpha_i = -\Lambda'.$$

Note that the second part of the proof could be carried out by referring to property 3) of Lyapunov transformations. □

LEMMA 3.5.3. *A linear system with constant coefficients is regular.*

PROOF. Consider a system $\dot{x} = Cx$, where $C$ is a constant matrix. Let

$$\lambda_1, \lambda_2, \dots, \lambda_n$$

be the eigenvalues of the matrix $C$; then

$$\{\operatorname{Re} \lambda_1, \operatorname{Re} \lambda_2, \dots, \operatorname{Re} \lambda_n\}$$

is the spectrum of the system. At the same time,

$$\Lambda' = \chi \left[ e^{-\int_{t_0}^{t} \operatorname{Sp} C\, d\tau} \right] = -\operatorname{Re} \operatorname{Sp} C = -\sum_{i=1}^{n} \operatorname{Re} \lambda_i. \quad \square$$

THEOREM 3.5.1. *A linear system reducible to a system with constant coefficients is regular.*

PROOF. Let there exist a Lyapunov transformation $x = L(t)y$ reducing system (3.5.1) to a system $\dot{y} = Cy$ with a constant matrix $C$. Consider normal fundamental matrices $X(t)$ and $Y(t)$ of both systems connected by the condition $X(t) = L(t)Y(t)$. Note that $\sigma_X = \sigma_Y$. According to the Ostrogradskiĭ-Liouville formula (1.0.5), we have

$$\operatorname{Det} X(t_0) e^{\int_{t_0}^{t} \operatorname{Sp} A(\tau)\, d\tau} = \operatorname{Det}(L(t) Y(t_0)) e^{\int_{t_0}^{t} \operatorname{Sp} C\, d\tau}.$$

Taking logarithms of the absolute values and dividing by $t$, we obtain

$$\frac{1}{t}\int_{t_0}^{t} \operatorname{Re}\operatorname{Sp} A(\tau)\, d\tau = \frac{1}{t}\ln|\operatorname{Det}(L(t)Y(t_0)X^{-1}(t_0))| + \frac{1}{t}\operatorname{Re}\operatorname{Sp} C(t - t_0).$$

The right-hand side of the last identity has a limit equal to $\operatorname{Re}\operatorname{Sp} C$ as $t \to \infty$; therefore, there exists

$$\lim_{t\to\infty} \frac{1}{t}\int_{t_0}^{t} \operatorname{Re}\operatorname{Sp} A(\tau)\, d\tau = \operatorname{Re}\operatorname{Sp} C = \sigma_Y.$$

Thus, the first condition of Lemma 3.5.1 is satisfied. The second condition is satisfied by virtue of the equality $\sigma_X = \sigma_Y$. □

Note additionally that the validity of the theorem also follows from Lemma 3.5.3, the symmetry of reducibility, and Lemma 3.5.2.

COROLLARY 3.5.1. *A linear system with periodic coefficients is regular.*

REMARK 3.5.1. The set of regular systems is wider than the set of systems reducible to systems with constant coefficients. Let us verify it by an example:

$$\dot{x} = x/(2\sqrt{t}), \qquad t \geqslant 1.$$

The general solution is $x = C\exp\sqrt{t}$. It follows from Lemma 3.5.1 that the equation is regular. However, $\exp\sqrt{t}$ cannot be represented as the product $L(t)\exp bt$, where $L(t)$ is Lyapunov, $b \in \mathbb{R}$. According to Erugin's Theorem 3.2.1, the equation is not reducible to an equation with constant coefficients.

THEOREM 3.5.2. *A linear system almost reducible to a system with constant coefficients is regular.*

The proof of this theorem is given in the Appendix, since it uses the stability of characteristic exponents and this notion will be introduced later.

## §6. Perron's regularity test. Coefficients of irregularity

Together with system (3.5.1) we shall consider the adjoint system

$$\dot{y} = -A^*(t)y. \tag{3.6.1}$$

Let

$$-\infty < \alpha_1 \leqslant \alpha_2 \leqslant \cdots \leqslant \alpha_n < \infty$$

be the spectrum of system (3.5.1) and let

$$\infty > \beta_1 \geqslant \beta_2 \geqslant \cdots \geqslant \beta_n > -\infty$$

be the spectrum of system (3.6.1).

THEOREM 3.6.1 (Perron). *System* (3.5.1) *is regular if and only if the complete spectra of system* (3.5.1) *and its adjoint are symmetric with respect to the origin, i.e.,*

$$\alpha_s + \beta_s = 0, \qquad s = 1, \dots, n. \tag{3.6.2}$$

PROOF. *Necessity.* Let system (3.5.1) be regular, let

$$X(t) = \{x_1(t), \dots, x_n(t)\}$$

be its normal basis, and let

$$\chi[x_s] = \alpha_s, \qquad s = 1, \dots, n.$$

Consider the basis

$$Y(t) = \{y_1(t), \dots, y_n(t)\}$$

of the adjoint system such that

$$Y^*(t)X(t) = E.$$

Such bases are said to be *reciprocal*. Let

$$\chi[y_s(t)] = \beta_s.$$

Since

$$(x_s, y_s) = 1, \qquad s = 1, \dots, n,$$

we have

$$\alpha_s + \beta_s \geqslant 0. \tag{3.6.3}$$

Let us prove the reverse inequality. In order to do this, we estimate the characteristic exponent of the $j$th element of the vector $y_s(t)$. Since $Y(t) = [X^{-1}(t)]^*$, we have

$$y_{js}(t) = \frac{\overline{X_{js}(t)}}{\operatorname{Det} X(t)} = \frac{\overline{X_{js}(t)}}{\operatorname{Det} X(t_0)} e^{-\int_{t_0}^t \overline{\operatorname{Sp} A(\tau)}\, d\tau}.$$

When calculating $\overline{X_{js}(t)}$ (the cofactor of the element $x_{js}(t)$ of the matrix $X(t)$), the $s$th column is omitted; therefore,

$$\chi[y_{js}] \leqslant \sum_{k=1}^n \alpha_k - \alpha_s - \sum_{k=1}^n \alpha_k.$$

Thus, $\chi[y_{js}] \leqslant -\alpha_s$, and this holds for any $j$; therefore,

$$\alpha_s + \beta_s \leqslant 0. \tag{3.6.4}$$

Comparing inequalities (3.6.3) and (3.6.4), we obtain

$$\alpha_s + \beta_s = 0, \qquad s = 1, \dots, n.$$

Now we verify that the basis $Y(t)$ under consideration is normal, i.e., its exponents constitute the complete spectrum of system (3.6.1). We check the validity of the Lyapunov equality (2.5.1) for $Y(t)$:

$$\begin{aligned}\sigma_Y &= \sum_{s=1}^n \beta_s = -\sum_{s=1}^n \alpha_s = -\chi\left[\exp\int_{t_0}^t \operatorname{Sp} A(\tau)\, d\tau\right] \\ &= -\lim_{t\to\infty} \frac{1}{t}\int_{t_0}^t \operatorname{Re}\operatorname{Sp} A(\tau)\, d\tau = \lim_{t\to\infty} \frac{1}{t}\int_{t_0}^t \operatorname{Re}\operatorname{Sp}(-A^*(\tau))\, d\tau.\end{aligned}$$

*Sufficiency.* Let the condition (3.6.2) hold. We verify that system (3.6.1) is regular. By the Lyapunov inequality,

$$\sum_{s=1}^{n} \alpha_s \geqslant \chi\left[\exp \int_{t_0}^{t} \operatorname{Sp} A(\tau)\, d\tau\right] = \Lambda,$$

$$\sum_{s=1}^{n} \beta_s \geqslant \chi\left[\exp \int_{t_0}^{t} \operatorname{Sp}(-A^*(\tau))\, d\tau\right] = \Lambda'.$$

After summing the inequalities, we obtain

$$\sum_{s=1}^{n} (\alpha_s + \beta_s) \geqslant \Lambda + \Lambda',$$

or, by (3.6.2),

$$0 \geqslant \Lambda + \Lambda'.$$

By the property of the characteristic exponent of a product, $0 \leqslant \Lambda + \Lambda'$; therefore, $\Lambda + \Lambda' = 0$, i.e., there exists

$$\lim_{t\to\infty} \frac{1}{t} \int_{t_0}^{t} \operatorname{Re} \operatorname{Sp} A(\tau)\, d\tau = S.$$

Thus, the first condition of Lemma 3.5.1 is satisfied.

Now we verify the second condition. Let, on the contrary,

$$\sum_{s=1}^{n} \alpha_s > S \qquad \text{and} \qquad \sum_{s=1}^{n} \beta_s > -S.$$

If one of these inequalities fails, the corresponding system would be regular, and then the adjoint system would also be regular, as follows from the proof of necessity. Summing these inequalities, we obtain

$$\sum_{s=1}^{n} (\alpha_s + \beta_s) > 0,$$

i.e., we come to a contradiction with (3.6.2). □

COROLLARY 3.6.1. *The system adjoint to a regular system is regular.*

COROLLARY 3.6.2. *The basis reciprocal to a normal basis of a regular system is itself normal. These bases are said to form a* binormal pair.

**Coefficients of irregularity.** When studying the influence of perturbations of a linear system on its properties it is essential to know whether the perturbed system is regular and if it is not, then to what extent. In order to do this, some numerical characteristics of deviation from regularity, which are called coefficients of irregularity, are introduced.

We note three of them.

1. Lyapunov's coefficient of irregularity ($\sigma$ or $\sigma_\Lambda$):

$$
\begin{aligned}
\sigma_\Lambda &= \sum_{s=1}^{n} \alpha_s + \chi\left[\exp \int_{t_0}^{t} - \operatorname{Sp} A(\tau)\, d\tau\right] \\
&= \sum_{s=1}^{n} \alpha_s + \Lambda' \qquad (3.6.5)\\
&= \sum_{s=1}^{n} \alpha_s - \underline{\lim}_{t\to\infty} \frac{1}{t}\int_{t_0}^{t} \operatorname{Re} \operatorname{Sp} A(\tau)\, d\tau.
\end{aligned}
$$

Note that

a) $\sigma_\Lambda \geqslant 0$ (this follows from inequality (3.5.3)),

b) $\sigma_\Lambda = 0$ if and only if the system is regular (this follows from Definition 3.5.1).

The second term in the definition of $\sigma_\Lambda$ is calculated straightforwardly from the coefficients of the system, and the first is estimated. We shall deal with estimates of characteristic exponents later on; according to the crudest estimate (3.2.5), we have

$$\sigma_\Lambda \leqslant n \overline{\lim_{t\to\infty}} \frac{1}{t}\int_{t_0}^{t} \|A(\tau)\|\, d\tau - \underline{\lim}_{t\to\infty} \frac{1}{t}\int_{t_0}^{t} \operatorname{Re} \operatorname{Sp} A(\tau)\, d\tau.$$

2. Perron's coefficient of irregularity ($\Pi$ or $\sigma_\Pi$). Let

$$-\infty < \alpha_1 \leqslant \alpha_2 \leqslant \cdots \leqslant \alpha_n < \infty$$

be the spectrum of the system $\dot{x} = A(t)x$ and let

$$\infty > \beta_1 \geqslant \beta_2 \geqslant \cdots \geqslant \beta_n > -\infty$$

be the spectrum of the adjoint system $\dot{y} = -A^*(t)y$. Then

$$\sigma_\Pi = \max(\alpha_s + \beta_s). \qquad (3.6.6)$$

Note that

a) $\sigma_\Pi \geqslant 0$. Let

$$X(t) = \{x_1(t), \dots, x_n(t)\}$$

be a basis of the first system and let

$$\chi[x_i] \leqslant \chi[x_{i+1}], \qquad i = 1, \dots, n-1.$$

The basis

$$Y(t) = [X^{-1}(t)]^* = \{y_1(t), \dots, y_n(t)\}$$

is the reciprocal basis and

$$\chi[y_i] = \beta_{s_i}.$$

Let us establish a one-to-one correspondence between the sequence $1, 2, \dots, l, \dots, n$ of indices of the characteristic exponents of the first basis and the analogous sequence $s_1, s_2, \dots, s_l, \dots, s_n$ for the second basis; we have

$$\alpha_i + \beta_{s_i} \geqslant 0, \qquad i = 1, \dots, n,$$

since $(x_i, y_i) = 1$.

Let us verify the inequality

$$\min_i(\alpha_i + \beta_{s_i}) \leqslant \min_i(\alpha_i + \beta_i).$$

This implies that $\sigma_\Pi \geqslant 0$. Let

$$\min_i(\alpha_i + \beta_i) = \alpha_l + \beta_l.$$

Obviously,

$$(*) \qquad \min_i(\alpha_i + \beta_{s_i}) \leqslant \alpha_l + \beta_{s_l}.$$

Consider two cases:
i. $l \leqslant s_l$. Then $\beta_l \geqslant \beta_{s_l}$, and from $(*)$ we obtain the inequality required.
ii. $l > s_l$. Then there exists a number $k$ such that $k < l \leqslant s_k$; therefore,

$$\min_i(\alpha_i + \beta_{s_i}) \leqslant \alpha_k + \beta_{s_k} \leqslant \alpha_l + \beta_l.$$

b) $\sigma_\Pi = 0$ if and only if the system is regular (Perron's Theorem 3.6.1).
3. Grobman's coefficient of irregularity ($\gamma$ or $\sigma_\Gamma$). Let

$$X(t) = \{x_1(t), \dots, x_n(t)\}$$

be some basis of system (3.5.1) and let

$$\chi[x_i] = \lambda_i, \qquad i = 1, \dots, n.$$

Here, to denote the exponents, we use the symbol $\lambda$ and not $\alpha$ in order to stress that the basis is not necessarily normal. Further, let

$$Y(t) - [X^{-1}(t)]^* = \{y_1(t), \dots, y_n(t)\}$$

be the reciprocal basis and let

$$\chi[y_i] = \mu_i, \qquad i = 1, \dots, n.$$

Denote by

$$\gamma' = \max(\lambda_i + \mu_i)$$

the *defect of reciprocal bases*; then

$$(3.6.7) \qquad \sigma_\Gamma \equiv \min_X \gamma'.$$

The coefficient of irregularity $\sigma_\Gamma$ is attained at normal bases **[9]**.
Note that
a) $\sigma_\Gamma \geqslant 0$, since $(x_i, y_i) = 1, i = 1, \dots, n$,
b) $\sigma_\Gamma = 0$ if and only if the system is regular.
The inequalities

$$0 \leqslant \sigma_\Pi \leqslant \sigma_\Gamma \leqslant n\sigma_\Pi, \qquad 0 \leqslant \sigma_\Gamma \leqslant \sigma_\Lambda \leqslant n\sigma_\Gamma$$

are valid **[9]**; thus, if one of the coefficients of irregularity vanishes, then the others also vanish.

EXAMPLE 3.6.1. Let us calculate the coefficients of irregularity for the system

$$\dot{x}_1 = (\sin \ln t + \cos \ln t)x_1,$$
$$\dot{x}_2 = 2(\sin \ln t + \cos \ln t)x_2.$$

For the fundamental matrix

$$X(t) = \begin{pmatrix} e^{t \sin \ln t} & 0 \\ 0 & e^{2t \sin \ln t} \end{pmatrix} = \{x_1(t), x_2(t)\}$$

we have

$$\chi[x_1] = \alpha_1 = 1, \qquad \chi[x_2] = \alpha_2 = 2.$$

Therefore, according to formula (3.6.5),

$$\sigma_\Lambda = 3 - \varliminf_{t\to\infty} \frac{1}{t} \int_{t_0}^{t} 3(\sin \ln \tau + \cos \ln \tau)\, d\tau$$
$$= 3 - \varliminf_{t\to\infty} 3 \sin \ln t = 3 - (-3) = 6.$$

For the reciprocal fundamental matrix

$$Y(t) = [X^{-1}(t)]^* = \begin{pmatrix} e^{-t \sin \ln t} & 0 \\ 0 & e^{-2t \sin \ln t} \end{pmatrix} = \{y_1(t), y_2(t)\}$$

we obtain

$$\chi[y_1] = \mu_1 = 1, \qquad \chi[y_2] = \mu_2 = 2.$$

Thus, according to (3.6.7),

$$\sigma_\Gamma = \max_{1,2}(\alpha_1 + \mu_1, \alpha_2 + \mu_2) = \max(2, 4) = 4.$$

Note further that $\beta_1 = 2$, $\beta_2 = 1$; hence, according to (3.6.6),

$$\sigma_\Pi = \max(1 + 2, 1 + 2) = 3.$$

## §7. The structure of fundamental matrices of a regular system. Generalized reducibility

In this section we give the results from Basov's paper **[3]**. Similar results were also obtained by Bogdanov **[6]** and Grobman **[17]**.

We introduce into consideration the class $A$ of $n \times n$ square matrices

$$Z(t) = \{z_{ij}(t)\}, \qquad i, j = 1, \dots, n,$$

such that

1. $z_{ij} \in C^1[t_0, \infty)$,
2. $\operatorname{Det} Z \neq 0, \quad t \in [t_0, \infty)$,
3. $\chi[z_{ij}] \leqslant 0$,
4. $\chi[\operatorname{Det} Z^{-1}] = 0$.

For example, any nonsingular matrix whose elements are polynomials belongs to the class $A$.

Note some properties of such matrices.

1. If $Z(t) \in A$, then $Z^{-1}(t) \in A$.

Indeed, according to the construction of the inverse matrix, we have

$$Z^{-1}(t) = \{v_{ij}(t)\} = \left\{ \frac{Z_{ji}(t)}{\operatorname{Det} Z(t)} \right\}, \qquad i, j = 1, \dots, n,$$

where $Z_{ji}$ is the cofactor of the element $z_{ji}$ of the matrix $Z(t)$. Hence,

$$v_{ij} \in C^1[t_0, \infty),$$

and, by Theorems 2.1.1 and 2.1.2,

$$\chi[v_{ij}] = \chi[Z_{ji} \operatorname{Det} Z^{-1}] \leqslant 0.$$

Moreover,

$$\operatorname{Det} Z(t) \operatorname{Det} Z^{-1}(t) = 1$$

and, by Theorem 2.1.2,

$$0 \leqslant \chi[\operatorname{Det} Z] + \chi[\operatorname{Det} Z^{-1}],$$

or, taking into account property 4 of matrices of the class $A$,

$$0 \leqslant \chi[\operatorname{Det} Z].$$

The reverse inequality $\chi[\operatorname{Det} Z] \leqslant 0$ follows from the process of calculation of the determinant.

2. If $Z_1(t) \in A$, $Z_2(t) \in A$, then $Z_1(t)Z_2(t) \in A$.

This statement can be easily verified straightforwardly.

3. If $Z(t) \in A$, then in any column and in any row of this matrix there is at least one element with zero characteristic exponent.

Indeed,

$$\operatorname{Det} Z = \sum z_{1a_1} z_{2a_2} \dots z_{na_n}$$

(where $a_1, a_2, \dots, a_n$ are different permutations of the numbers $1, 2, \dots, n$). At the same time $\chi[\operatorname{Det} Z] = 0$; hence, according to Theorem 2.1.1 on the right-hand side of the last equality there is at least one term with zero characteristic exponent; therefore, all the cofactors constituting this term have zero characteristic exponent, and they are from different rows and columns of the matrix $Z(t)$.

THEOREM 3.7.1. *A system $\dot{x} = A(t)x$ (see (3.5.1)) is regular if and only if it has a normal fundamental matrix $X(t)$ of the form*

$$X(t) = Z(t)e^{\operatorname{diag}[\alpha_1, \dots, \alpha_n]t}, \tag{3.7.1}$$

*where $Z(t) \in A$, $\alpha_i \in \mathbb{R}$ are finite, $i = 1, \dots, n$.*

PROOF. *Necessity.* Let

$$X(t) = \{x_1(t), \dots, x_n(t)\}$$

be a normal fundamental matrix of system (3.5.1) and let

$$\chi[x_i] = \alpha_i, \qquad i = 1, \dots, n.$$

Rewrite it in the form

$$X(t) = X(t)e^{-\operatorname{diag}[\alpha_1,\dots,\alpha_n]t}e^{\operatorname{diag}[\alpha_1,\dots,\alpha_n]t} = Z(t)e^{\operatorname{diag}[\alpha_1,\dots,\alpha_n]t}. \tag{3.7.2}$$

Let us show that $Z(t) \in A$. Indeed, according to (3.7.2) we have

1. $Z \in C^1[t_0, \infty)$,
2. $\operatorname{Det} Z(t) = \operatorname{Det} X(t)e^{-\sum_{i=1}^n \alpha_i t} \neq 0$,
3. $Z(t) = \{x_1(t)e^{-\alpha_1 t}, x_2(t)e^{-\alpha_2 t}, \dots, x_n(t)e^{-\alpha_n t}\}$,

i.e.,

$$\chi[z_{ij}] \leqslant 0, \qquad i, j, = 1, \dots, n,$$

4.

$$\operatorname{Det} Z^{-1}(t) = \operatorname{Det} e^{\operatorname{diag}[\alpha_1,\dots,\alpha_n]t} \operatorname{Det} X^{-1}(t) = e^{t\sum_{i=1}^n \alpha_i} \exp\left[-\int_{t_0}^t \operatorname{Sp} A(\tau)\,d\tau\right] \frac{1}{\operatorname{Det} X(t_0)},$$

i.e.,

$$\chi[\operatorname{Det} Z^{-1}] = \sum_{i=1}^n \alpha_i + \Lambda' = \sigma_\Lambda. \tag{3.7.3}$$

Now we use the regularity of the system: $\sigma_\Lambda = 0$; hence, $\chi[\operatorname{Det} Z^{-1}] = 0$.

*Sufficiency.* Let (3.7.1) hold. Let us show that the system is regular.

We have

$$\begin{aligned} X(t) &= Z(t) \exp \operatorname{diag}[\alpha_1, \dots, \alpha_n]t \\ &= \{z_1(t) \exp \alpha_1 t, z_2(t) \exp \alpha_2 t, \dots, z_n(t) \exp \alpha_n t\} \\ &= \{x_1(t), x_2(t), \dots, x_n(t)\}. \end{aligned}$$

Since each vector $z_i(t)$ has at least one element with zero characteristic exponent and the other elements have nonpositive exponents, we have

$$\chi[X_i] = \alpha_i, \qquad i = 1, \dots, n.$$

Moreover, $\chi[\operatorname{Det} Z^{-1}] = 0$; therefore, by relation (3.7.3), $\sigma_\Lambda = 0$, i.e., the system is regular and the fundamental matrix is normal. □

COROLLARY 3.7.1. *If system* (3.5.1) *has a fundamental matrix of the form* (3.7.1), *then the system is regular, the fundamental matrix is normal, and*

$$\{\alpha_1, \alpha_2, \dots, \alpha_n\}$$

*is the spectrum of the system.*

The validity of the corollary follows from the proof of sufficiency.

The theorem proven allows us to extend the notion of regularity to systems whose coefficients are not bounded, but satisfy the requirement that the characteristic exponents be finite. A restriction so severe on the coefficients was initially introduced by Lyapunov in order to guarantee that the characteristic exponents of the solutions and the function

$$\int_{t_0}^t \operatorname{Sp} A(\tau)\,d\tau$$

are finite. Whenever this restriction can be dropped, it should.

DEFINITION 3.7.1. A system

$$\dot{x} = A(t)x, \qquad A \in C[t_0, \infty),$$

is said to be *regular* if it has a fundamental matrix $X(t)$ of the form

$$X(t) = Z(t)e^{\operatorname{diag}[\alpha_1, \alpha_2, \ldots, \alpha_n]t},$$

where

$$Z(t) \in A, \qquad \alpha_i \in \mathbb{R} \quad \text{are finite}, \qquad i = 1, \ldots, n.$$

EXAMPLE 3.7.1. We give an example of a regular system with unbounded coefficients, $\dot{x} = -x$, $\dot{y} = tx + y$; here

$$X(t) = \begin{pmatrix} e^{-t} & 0 \\ \frac{e^{-t}(2t+1)}{-4} & e^t \end{pmatrix} = \begin{pmatrix} 1 & 0 \\ -\frac{2t+1}{4} & 1 \end{pmatrix} \begin{pmatrix} e^{-t} & 0 \\ 0 & e^t \end{pmatrix},$$

i.e., the fundamental matrix has the form (3.7.1).

Naturally, all the available results on regular systems can be obtained starting with Definition 3.7.1, when the proofs turn out to be simpler for many of them. Let us give an example.

PERRON'S THEOREM 3.6.1 (necessity). *If a system $\dot{x} = A(t)x$ is regular, then its spectrum and the spectrum of the adjoint system are symmetric with respect to the origin.*

PROOF. There exists a fundamental matrix $X(t)$ of the form (3.7.1),

$$X(t) = Z(t) \exp \operatorname{diag}[\alpha_1, \ldots, \alpha_n]t.$$

The reciprocal matrix is

$$Y(t) = [X^{-1}(t)]^* = [Z^{-1}(t)]^* e^{-\operatorname{diag}[\alpha_1, \ldots, \alpha_n]t}.$$

By Corollary 3.7.1,

$$\{\alpha_1, \alpha_2, \ldots, \alpha_n\}$$

is the spectrum of the initial system, and, since $[Z^{-1}(t)]^* \in A$, we have that

$$\{-\alpha_1, \ldots, -\alpha_n\}$$

is the spectrum of the adjoint system. □

DEFINITION 3.7.2. A transformation $x = U(t)y$ is called a *generalized Lyapunov transformation* if the matrix $U(t) \in A$.

THEOREM 3.7.2. *A system*

$$\dot{x} = A(t)x, \qquad A \in C[t_0, \infty),$$

*is regular if and only if it is reducible in the generalized sense to a diagonal system with real coefficients.*

PROOF. *Necessity.* Let the system be regular. Let us take a fundamental matrix of the form

$$X(t) = Z(t) \exp \operatorname{diag}[\alpha_1, \dots, \alpha_n]t$$

and subject the system to the transformation

$$x = X(t) \exp(-\operatorname{diag}[\alpha_1, \dots, \alpha_n]t)y \equiv U(t)y.$$

We obtain

$$\dot{y} = (U^{-1}AU - U^{-1}\dot{U})y = B(t)y. \tag{3.7.4}$$

Therefore,

$$\begin{aligned} B(t) &= e^{\operatorname{diag}[\alpha_1,\dots,\alpha_n]t} X^{-1}(t)A(t)X(t)e^{-\operatorname{diag}[\alpha_1,\dots,\alpha_n]t} \\ &\quad - e^{\operatorname{diag}[\alpha_1,\dots,\alpha_n]t} X^{-1}(t)\Big[A(t)X(t)e^{-\operatorname{diag}[\alpha_1,\dots,\alpha_n]t} \\ &\qquad\qquad - X(t)e^{-\operatorname{diag}[\alpha_1,\dots,\alpha_n]t} \operatorname{diag}[\alpha_1, \dots, a_n]\Big] \\ &= \operatorname{diag}[\alpha_1, \dots, \alpha_n]. \end{aligned}$$

*Sufficiency.* Let the transformation $x = U(t)y$, where $U(t) \in A$, reduce the initial system to system (3.7.4), where

$$B = \operatorname{diag}[\alpha_1, \dots, \alpha_n].$$

One of the fundamental matrices of this system is the matrix

$$Y(t) = \exp \operatorname{diag}[\alpha_1, \dots, \alpha_n]t.$$

Then, by virtue of our transformation, the initial system has a fundamental matrix

$$X(t) = U(t) \exp \operatorname{diag}[\alpha_1, \dots, \alpha_n]t,$$

and, by Definition 3.7.1, this system is regular. □

## §8. Regularity of a triangular system

Consider a system

$$\begin{aligned} \dot{x}_1 &= a_{11}(t)x_1, \\ \dot{x}_2 &= a_{21}(t)x_1 + a_{22}(t)x_2, \\ &\dots\dots\dots\dots \\ \dot{x}_n &= a_{n1}(t)x_1 + a_{n2}(t)x_2 + \dots + a_{nn}(t)x_n, \end{aligned} \tag{3.8.1}$$

where

$$a_{ij} \in C[t_0, \infty), \qquad i, j = 1, \dots, n, \quad j \leqslant i, \quad \sup|a_{ij}(t)| \leqslant M.$$

THEOREM 3.8.1 (Lyapunov). *The triangular system* (3.8.1) *with real diagonal is regular if and only if its diagonal coefficients have finite mean values, i.e.,*

$$\mu_k = \lim_{t\to\infty} \frac{1}{t} \int_{t_0}^{t} a_{kk}(u)\, du, \qquad k = 1, \dots, n. \tag{3.8.2}$$

PROOF. *Necessity.* Let system (3.8.1) be regular. Let us show that equalities (3.8.2) hold. We introduce the notations

$$\overline{\mu_k} = \overline{\lim_{t\to\infty}} \frac{1}{t} \int_{t_0}^{t} a_{kk}(u)\,du,$$

$$\underline{\mu_k} = \underline{\lim_{t\to\infty}} \frac{1}{t} \int_{t_0}^{t} a_{kk}(u)\,du,$$

$$A_k = \exp \int_{t_0}^{t} a_{kk}(u)\,du.$$

Successively integrating (3.8.1), we obtain the lower triangular fundamental matrix

$$X(t, t_0) = \{x_1(t, t_0), \dots, x_n(t, t_0)\}.$$

Let

$$x_k(t, t_0) = (x_{k1}(t, t_0), x_{k2}(t, t_0), \dots, x_{kn}(t, t_0))^\top;$$

then

$$\begin{aligned} x_{kj}(t, t_0) &= 0 && \text{for} \quad k > j,\\ x_{kk}(t, t_0) &= A_k && \text{for} \quad k = j,\\ x_{kj}(t, t_0) &= A_j \int_{t_0}^{t} A_j^{-1} \sum_{s=k}^{j-1} a_{js} x_{ks}(\tau, t_0)\,d\tau && \text{for} \quad k < j. \end{aligned}$$

Therefore, the fundamental matrix constructed may not be normal. By Theorem 2.4.2, there exists a lower triangular constant matrix $C$ with diagonal elements equal to unity such that

$$X(t) = X(t, t_0)C = \{x_1(t), \dots, x_n(t)\}$$

is already normal and $x_{kk}(t) = A_k$; the matrix $X(t)$ is also lower triangular. Since

$$\chi[x_{kk}] = \overline{\mu_k},$$

we have

$$\chi[x_k] = \alpha_k \geqslant \overline{\mu_k}.$$

For the binormal (upper triangular) matrix

$$Y(t) = \left[X^{-1}(t)\right]^* = \{y_1(t), \dots, y_n(t)\}$$

we have $y_{kk}(t) = A_k^{-1}$; therefore,

$$\chi[y_{kk}] = -\underline{\mu_k}, \qquad \chi[y_k] = \beta_k \geqslant -\underline{\mu_k}.$$

The inequality $0 \leqslant \overline{\mu_k} - \underline{\mu_k}$ always holds and, by virtue of the regularity of system (3.8.1), we have

$$\alpha_k + \beta_k = 0.$$

Combining these inequalities, we obtain

$$0 = \alpha_k + \beta_k \geqslant \overline{\mu_k} - \underline{\mu_k} \geqslant 0;$$

therefore,

$$\overline{\mu_k} = \underline{\mu_k},$$

and (3.8.2) is proved.

*Sufficiency.* Let (3.8.2) hold. Let us show that system (3.8.1) is regular. Consider the matrix

$$Z(t) = \{z_1(t), \dots, z_n(t)\},$$

where

$$z_k(t) = \left(z_{k1}(t), z_{k2}(t), \dots, z_{kn}(t)\right)^\top$$

and

$$\begin{aligned} z_{kj}(t) &= 0 && \text{for} \quad j < k, \\ z_{kk}(t) &= A_k && \text{for} \quad j = k, \\ z_{kj}(t) &= A_j \int_{\tau j}^{t} A_j^{-1} \sum_{s=k}^{j-1} a_{js}(\tau) z_{ks}(\tau)\, d\tau && \text{for} \quad j > k, \end{aligned}$$

where

$$\tau_j = t_0 \quad \text{if} \quad \mu_j \leqslant \mu_k$$

and

$$\tau_j = \infty \quad \text{if} \quad \mu_j > \mu_k.$$

Let us show that $\chi[z_k] = \mu_k$. Indeed,

$$\begin{aligned} \chi[z_{kk}] &= \mu_k, \\ \chi[z_{kk+1}] &= \chi\left[A_{k+1} \int_{\tau_{k+1}}^{t} A_{k+1}^{-1} a_{k+1k}(\tau) z_{kk}(\tau)\, d\tau\right] \\ &\leqslant \mu_{k+1} - \mu_{k+1} + \mu_k + \chi[a_{k+1k}] \leqslant \mu_k. \end{aligned}$$

In the derivation of this estimate we have used Theorem 2.1.5. Further we reason by induction. Let

$$\chi[z_{ki}] \leqslant \mu_k \qquad \text{for} \qquad i = k, \dots, j-1.$$

Let us estimate

$$\chi[z_{kj}] \leqslant \mu_j - \mu_j + \max_s \{\chi[a_{js}]\} + \max_s \chi[z_{ks}] \leqslant \mu_k.$$

Hence, by definition,

$$\chi[z_k] = \mu_k.$$

Let us show that $z_k(t)$ is a solution of system (3.8.1). Compare $z_k(t)$ with $x_k(t, t_0)$ elementwise. If $\tau_j = t_0$, then the coincidence is complete; if $\tau_j = \infty$, then the exponent of the integrand is negative and, rewriting the integral in the form

$$\int_{t_0}^{t} = \int_{t_0}^{\infty} + \int_{\infty}^{t},$$

we see that the first integral on the right-hand side of the last equality converges and $z_{kj}(t)$ differs from $x_{kj}(t, t_0)$ by the additive term $cA_j$; thus, we still are in the set of solutions. Hence,

$$z_1(t), z_2(t), \dots, z_n(t)$$

are solutions of system (3.8.1); moreover, they are linearly independent, since

$$\operatorname{Det} Z(t_0) = 1.$$

As proved above,

$$\sum_{k=1}^{n} \chi[z_k] = \sum_{k=1}^{n} \mu_k,$$

and, according to (3.8.2), there exists

$$\lim_{t\to\infty} \frac{1}{t} \int_{t_0}^{t} \operatorname{Sp} A(\tau)\, d\tau = \sum_{k=1}^{n} \mu_k.$$

Therefore, the conditions of Lemma 3.5.1 are satisfied, system (3.8.1) is regular, and the matrix $Z(t)$ constructed is normal fundamental. □

COROLLARY 3.8.1. *If the diagonal elements of system* (3.8.1) *have finite mean values, then these values give the complete spectrum of the system and the system itself is regular.*

REMARK 3.8.1. Using the methods from the previous section, we can weaken the conditions for the coefficients of the system and require only that the characteristic exponents of the off-diagonal coefficients be nonpositive.

## §9. Properties of solutions of regular systems

In this section we present the results of Vinograd **[15]**.

THEOREM 3.9.1. *Any nontrivial solution $x(t)$ of a regular system* (3.5.1) *has a sharp characteristic exponent, i.e., there exists the limit*

$$\lim_{t\to\infty} \frac{1}{t} \ln \|x(t)\| = \chi[x]. \tag{3.9.1}$$

PROOF. Let us take a basis $X(t)$ of system (3.5.1) with the solution $x(t)$ in its first column. Using this basis, let us construct the Perron transformation $x = U(t)y$ (Theorem 3.3.1); applying it, we obtain the triangular system $\dot{y} = B(t)y$ which is also regular (Lemma 3.5.2). According to Theorem 3.8.1, there exists

$$\begin{aligned}\lim_{t\to\infty} \frac{1}{t} \int_{t_0}^{t} b_{11}(\tau)\, d\tau &= \lim_{t\to\infty} \frac{1}{t} \int_{t_0}^{t} d \ln \|x(\tau)\| \\ &= \lim_{t\to\infty} \frac{1}{t} \ln \|x(t)\|;\end{aligned}$$

hence, (3.9.1) is valid. Recall that the equality

$$b_{11} = \frac{d \ln \|x(t)\|}{dt}$$

was proved in Theorem 3.3.1. □

To present the following result we recall the process of orthogonalization of the basis

$$X(t) = \{x_1(t), \dots, x_n(t)\}$$

described in Lemma 3.3.1. The notation is preserved. Thus,

$$\begin{aligned}\xi_1 = x_1, \qquad e_1 = \xi_1/\|\xi_1\|,\\ \xi_2 = x_2 - (x_2, e_1)e_1,\end{aligned}$$

or $x_2 = \xi_2 + x_2^*$, where $x_2^*$ belongs to the subspace spanned by the vector $\xi_1$, and $\xi_1$ and $\xi_2$ are orthogonal by construction. For vectors $x \in \mathbb{C}^n$ and $y \in \mathbb{C}^n$ we set

$$\cos^2(x, y) = \frac{|(x, y)|^2}{\|x\|^2\|y\|^2}, \qquad \sin(x, y) = (1 - \cos^2(x, y))^{1/2}.$$

Then, if $a$ and $b$ are orthogonal and $x = a + b$, we have

$$\|a\| = \|x\| \sin(x, b).$$

In these terms we have

$$\|\xi_2\| = \|x_2\| \sin(x_2, x_2^*).$$

Following the process of orthogonalization , we obtain

$$x_k = \xi_k + x_k^*, \qquad k \geqslant 2,$$

where $x_k^*$ belongs to the subspace spanned by the vectors $x_1, x_2, \dots, x_{k-1}$, and, therefore, is orthogonal to $\xi_k$. Thus,

$$\|\xi_k\| = \|x_k\| \sin(x_k, x_k^*), \qquad k \geqslant 2. \tag{3.9.2}$$

The vector $x_k^*$ is the projection of $x_k$ on the subspace with the basis $x_1, \dots, x_{k-1}$ and the "angle" $(x_k, x_k^*)$ is the "angle" between $x_k$ and the subspace indicated. If the basis $X(t)$ is real, then these terms have a direct geometric interpretation.

THEOREM 3.9.2. *For any normal basis*

$$X(t) = \{x_1(t), \dots, x_n(t)\}$$

*of a regular system the following limits exist and are equal to zero*:

$$\lim_{t\to\infty} \frac{1}{t} \ln \sin(x_k, x_k^*) = 0, \qquad k = 2, 3, \dots, n. \tag{3.9.3}$$

PROOF. According to Lemma 3.3.1, in its notation we have

$$\begin{aligned} |\operatorname{Det} X(t)| &= |\operatorname{Det} U(t) \operatorname{Det} R(t)| = \operatorname{Det} R(t) \\ &= \prod_{k=1}^{n} \|\xi_k\| \overset{(3.9.2)}{=} \|x_1\| \prod_{k=2}^{n} \|x_k\| \sin(x_k, x_k^*). \end{aligned}$$

At the same time, by virtue of the notation from (3.5.2), we have

$$\begin{aligned} \Lambda' &= -\varliminf_{t\to\infty} \frac{1}{t} \int_{t_0}^{t} \operatorname{Re} \operatorname{Sp} A(\tau)\, d\tau \\ &= -\varliminf_{t\to\infty} \frac{1}{t} \ln |\operatorname{Det} X(t)| \\ &= -\varliminf_{t\to\infty} \frac{1}{t} \ln \left( \prod_{k=1}^{n} \|x_k\| \prod_{k=2}^{n} \sin(x_k, x_k^*) \right) \\ &\overset{(3.9.1)}{=} -\lim_{t\to\infty} \frac{1}{t} \ln \left( \prod_{k=1}^{n} \|x_k\| \right) - \varliminf_{t\to\infty} \frac{1}{t} \ln \prod_{k=2}^{n} \sin(x_k, x_k^*). \end{aligned} \tag{3.9.4}$$

Since the system under consideration is regular, we have

$$\Lambda' + \lim_{t\to\infty} \frac{1}{t} \ln \left( \prod_{k=1}^{n} \|x_k\| \right) = 0;$$

therefore,

$$\underline{\lim}_{t\to\infty} \frac{1}{t} \ln \prod_{k=2}^{n} \sin(x_k, x_k^*) = 0.$$

Let us write the following chain of inequalities which proves that (3.9.3) is valid:

$$\begin{aligned} 0 &= \underline{\lim}_{t\to\infty} \frac{1}{t} \ln \prod_{k=2}^{n} \sin(x_k, x_k^*) \\ &\leqslant \underline{\lim}_{t\to\infty} \frac{1}{t} \ln \sin(x_k, x_k^*) \\ &\leqslant \overline{\lim}_{t\to\infty} \frac{1}{t} \ln \sin(x_k, x_k^*) \\ &\leqslant \chi[1] = 0. \quad \square \end{aligned}$$

REMARK 3.9.1. The normality of the basis in Theorem 3.9.2 is essential. Consider an example:

$$\dot{x} = x, \qquad \dot{y} = -y.$$

For the basis

$$X(t) = \{x_1(t), x_2(t)\} = \begin{pmatrix} e^t, & e^t \\ e^{-t}, & 0 \end{pmatrix}$$

we have

$$\begin{aligned} \chi[\sin(x_1, x_2)] &= \chi\left(1 - \frac{e^{4t}}{e^{2t}(e^{2t} + e^{-2t})}\right)^{1/2} \\ &= \chi\left[\frac{1}{(e^{4t}+1)^{1/2}}\right] \\ &= -2. \end{aligned}$$

The system is regular, but the basis $X(t)$ is not normal; because of this, the condition (3.9.3) is violated.

The necessary criteria of regularity are at the same time sufficient.

THEOREM 3.9.3. *If some basis*

$$X(t) = \{x_1(t), \dots, x_n(t)\}$$

*of system* (3.5.1) *has sharp characteristic exponents and the functions*

$$\sin(x_k; x_k^*), \qquad k = 2, \dots, n,$$

*also have sharp characteristic exponents that, moreover, are equal to zero, then the basis is normal and the system is regular.*

PROOF. Let

$$\chi[x_k] = \alpha_k, \qquad k = 1, 2, \ldots, n.$$

According to relations (3.5.3) and (3.9.4), we have

$$\sum_{k=1}^{n} \alpha_k \geqslant \Lambda \geqslant -\Lambda' = \lim_{t\to\infty} \frac{1}{t} \ln \prod_{k=1}^{n} \|x_k\| + \lim_{t\to\infty} \frac{1}{t} \ln \prod_{k=2}^{n} \sin(x_k, x_k^*) = \sum_{k=1}^{n} \alpha_k.$$

Thus,

$$\sum_{k=1}^{n} \alpha_k = \Lambda = -\Lambda',$$

i.e., the basis is normal and system (3.5.1), according to Definition 3.1.1, is regular. □

Note that the following regularity test was proven in **[9]**.

THEOREM 3.9.4. *System* (3.5.1) *is regular if and only if the Gram determinants consisting of any of its solutions have sharp characteristic exponents.*

CHAPTER IV

# Stability and Small Perturbations of the Coefficients of Linear Systems

In the first section of this chapter we consider the properties of stable and asymptotically stable linear systems, formulate the corresponding results for autonomous, periodic, and Lappo-Danilevskiĭ systems, and also study linear perturbations of autonomous systems preserving stability and asymptotic stability.

In what follows it will be shown that it is impossible to transfer literally the results on perturbations to the case of variable coefficients; this will lead us to the consideration of uniform stability and uniform asymptotic stability. For systems with these properties we shall consider admissible linear perturbations.

At the end of the chapter we indicate the variations of coefficients of linear systems preserving their spectrum and give methods of estimating the growth of solutions via the coefficients of a system.

## §1. On stability of linear systems

In this section we dwell on the Lyapunov stability of solutions of linear systems. Recall the main definitions.

Let a normal system

$$\dot{x} = f(t, x), \qquad x \in \mathbb{C}^n,$$

be such that the conditions of existence and uniqueness of the solution of the Cauchy problem are satisfied in some domain

$$G \subset \mathbb{R} \times \mathbb{C}^n,$$

and the solution $x(t, t_0, x_0)$ can be extended to $[t_0, \infty)$. Here

$$x(t_0, t_0, x_0) = x_0.$$

We call this solution *unperturbed*, the solution $x(t, t_0, \eta)$ *perturbed*, and the difference $x_0 - \eta$ the *perturbation*.

DEFINITION 4.1.1. The solution $x(t, t_0, x_0)$ is said to be *Lyapunov stable* if for any $\varepsilon > 0$ there exists a $\delta > 0$ such that the condition

$$\|x_0 - \eta\| < \delta$$

implies

$$\|x(t, t_0, x_0) - x(t, t_0, \eta)\| < \varepsilon$$

for all $t \geqslant t_0$.

DEFINITION 4.1.2. The solution $x(t, t_0, x_0)$ is called *unstable* if it is not stable, i.e., if there exists an $\varepsilon > 0$ such that for any $\delta > 0$ there is an $\eta_1 \in \mathbb{C}^n$ satisfying the condition

$$\|x_0 - \eta_1\| < \delta$$

and a moment of time $t_1 > t_0$ such that

$$\|x(t_1, t_0, x_0) - x(t_1, t_0, \eta_1)\| \geqslant \varepsilon.$$

DEFINITION 4.1.3. The solution $x(t, t_0, x_0)$ is said to be *asymptotically stable* if
1) it is Lyapunov stable,
2) there exists a $\Delta > 0$ such that the condition $\|x_0 - \eta\| < \Delta$ implies

$$\lim_{t\to\infty} \|x(t, t_0, x_0) - x(t, t_0, \eta)\| = 0.$$

Note that the ball

$$\|x_0 - x\| < \Delta$$

is called the *domain of attraction* of the solution $x(t, t_0, x_0)$.

DEFINITION 4.1.4. The solution $x(t, t_0, x_0)$ is said to be *globally asymptotically stable* if, in the previous definition, $\Delta = \infty$.

We omit the explanations of the definitions given above since the reader can find this material in, e.g., **[25]** or **[19]**. Nevertheless, we note two facts. First, stability or instability do not depend on the specific choice of $t_0$ since in any finite interval the closeness of solutions under a small perturbation is guaranteed by the theorem on integral continuity **[5]**; second, the Lyapunov stability is nothing else but the uniform continuity of the solution $x(t, t_0, x_0)$ on $[t_0, \infty)$ with respect to the initial vector $x_0$.

Now we pass to linear systems. Together with a linear nonhomogeneous system

$$\dot{x} = A(t)x + f(t), \tag{4.1.1}$$

where

$$x \in \mathbb{C}^n, \qquad f, A \in C[t_0, \infty), \qquad \sup_{t\geqslant t_0} \|A(t)\| < \infty,$$

we shall consider the corresponding homogeneous system

$$\dot{x} = A(t)x. \tag{4.1.2}$$

We verify that the stability of any solution of these systems is equivalent to the stability of the trivial solution of the homogeneous system. Indeed, let $X(t, t_0)$ be the fundamental matrix of system (4.1.2),

$$X(t_0, t_0) = E;$$

then the solution $x(t, t_0, x_0)$ of system (4.1.1) is written as

$$x(t, t_0, x_0) = X(t, t_0)x_0 + \int_{t_0}^{t} X(t, \tau)f(\tau)\, d\tau. \tag{4.1.3}$$

If $f(t) \equiv 0$ for $t \geqslant t_0$, then the second term vanishes, and (4.1.3) in this case represents a solution of the homogeneous system. Consider the difference of the perturbed and unperturbed solutions,

$$x(t, t_0, x_0) - x(t, t_0, \eta) = X(t, t_0)(x_0 - \eta);$$

the difference

$$z(t) = x(t, t_0, x_0) - x(t, t_0, \eta)$$

for any $f(t)$ (and also for $f(t) \equiv 0$) is the solution of the homogeneous system (4.1.2) with the initial condition $z(t_0) = x_0 - \eta$.

Therefore, the problems of stability of the unperturbed solution $x(t, t_0, x_0)$ either of system (4.1.1) or (4.1.2) are reduced to the following: whether for any $\varepsilon > 0$ there exists a $\delta > 0$ such that the condition

$$\|z(t_0)\| < \delta$$

implies the condition

$$\|z(t)\| < \varepsilon$$

for all $t \geqslant t_0$. Actually, this is the problem of stability of the trivial solution of the homogeneous system (4.1.2).

Our considerations lead to the following criterion: if some solution of a linear homogeneous or nonhomogeneous system is stable, then the trivial solution of the homogeneous system is stable and, conversely, the stability of the latter implies the stability of any solution of both the homogeneous and the nonhomogeneous linear systems, and this fact does not depend on the specific form of $f(t)$.

Analogous results hold for asymptotic stability.

Thus, all the solutions of systems (4.1.1) and (4.1.2) are stable, asymptotically stable, or unstable simultaneously. This is an important property of linear systems. That is why the following terminology is standard for them: a *stable linear system*, an *unstable linear system*, an *asymptotically stable linear system*.

REMARK 4.1.1. Generally speaking, nonlinear systems do not have the property mentioned above.

EXAMPLE 4.1.1. $\dot{x} = \cos x$. This equation has the solutions

$$x_k = \pi/2 + k\pi, \qquad k \in \mathbb{Z}.$$

In this set asymptotically stable ($k = 2m$) and unstable ($k = 2m + 1$) solutions alternate, which can be easily seen from the graphs of the integral curves shown in Figure 1 on the next page.

The arguments given above easily establish the validity of the following theorem.

THEOREM 4.1.1. *A linear nonhomogeneous system is stable (asymptotically stable) if and only if the trivial solution of the linear homogeneous system is stable (asymptotically stable).*

COROLLARY 4.1.1. *A linear nonhomogeneous system is stable (asymptotically stable) if and only if the corresponding homogeneous system is stable (asymptotically stable).*

Now we pass to the study of homogeneous systems.

Consider system (4.1.2),

$$\dot{x} = A(t)x,$$

where

$$x \in \mathbb{C}^n, \qquad A \in C[t_0, \infty), \qquad \sup_{t \geqslant t_0} \|A(t)\| < \infty.$$

THEOREM 4.1.2. *A linear homogeneous system is stable if and only if each of its solutions $x(t)$ is bounded for $t \geqslant t_0$.*

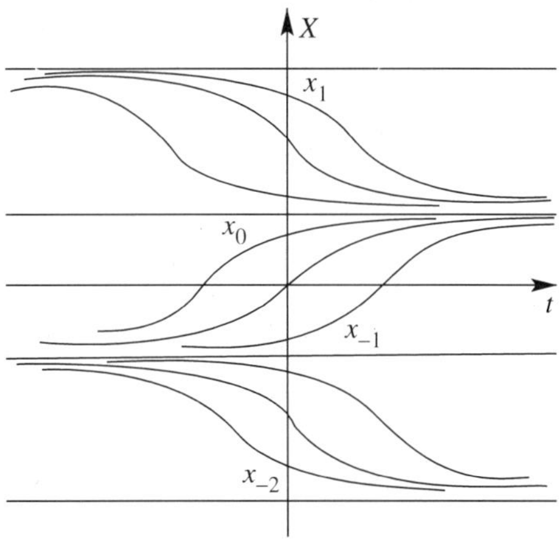


FIGURE 1

PROOF. We show that the stability of the system implies the boundedness of any of its solutions. Let, conversely, there exist a solution $z(t)$ of system (4.1.2) unbounded for $[t_0, \infty)$. Obviously,

$$\|z(t_0)\| \neq 0.$$

Fix $\varepsilon > 0$; for it determine $\delta > 0$, whose existence is guaranteed by the stability of the trivial solution, and form the solution

$$x(t) = \frac{z(t)}{\|z(t_0)\|} \cdot \frac{\delta}{2}.$$

Since

$$\|x(t_0)\| = \delta/2 < \delta,$$

by virtue of stability, we have

$$\|x(t)\| < \varepsilon \qquad \text{for} \qquad t \geqslant t_0;$$

this contradicts our assumption that $z(t)$ is unbounded.

Let us carry out the considerations in the opposite direction. Suppose $X(t, t_0)$ is the fundamental matrix of (4.1.2), its columns are solutions of the system, and they are bounded and $n$ in number; therefore, there exists a constant $C$ such that

$$\|X(t, t_0)\| \leqslant C \qquad \text{for} \qquad t \geqslant t_0.$$

Any solution $x(t)$ is written in the form

$$x(t) = X(t, t_0)x(t_0).$$

Hence,

$$\|x(t)\| \leqslant \|X(t, t_0)\| \cdot \|x(t_0)\| \leqslant C\|x(t_0)\|.$$

Take $\varepsilon > 0$ and determine $\delta = \varepsilon/C$ according to it. Obviously, the inequality

$$\|x(t)\| < \varepsilon \qquad \text{for} \qquad t \geqslant t_0$$

follows from the inequality

$$\|x(t_0)\| < \delta.$$

Thus, the trivial solution is stable and system (4.1.2) is also stable. □

COROLLARY 4.1.2. *All the solutions of a stable linear homogeneous system are simultaneously either bounded or unbounded for* $t \geqslant t_0$.

Indeed, the general solution of system (4.1.1) is written in the form

$$x(t) = X(t)c + x_1(t),$$

where $X(t)$ is a fundamental matrix of system (4.1.2), $c \in \mathbb{C}^n$, and $x_1(t)$ is a particular solution of system (4.1.1). Only the latter affects the boundedness, since the matrix $X(t)$ is bounded.

EXAMPLE 4.1.2. Consider two nonhomogeneous asymptotically stable equations:

$$\begin{aligned} &1. \qquad \dot{x} = -x + e^t, \qquad && x(t) = ce^{-t} + e^t/2, \\ &2. \qquad \dot{x} = -x + e^{-t}, \qquad && x(t) = ce^{-t} + te^{-t}. \end{aligned}$$

In the first example all the solutions are unbounded, in the second, bounded.

REMARK 4.1.2. Example 4.1.1 shows that in the case of nonlinear systems there is no correlation between stability of solutions and their boundedness.

THEOREM 4.1.3. *A linear homogeneous system is asymptotically stable if and only if all its solutions* $x(t)$ *tend to zero as* $t \to \infty$, *i.e.*,

$$\lim_{t\to\infty} \|x(t)\| = 0.$$

PROOF. Let the trivial solution be asymptotically stable, i.e., let there exist $\Delta > 0$ such that the inequality

$$\|x(t_0)\| < \Delta$$

implies

$$\lim_{t\to\infty} \|x(t)\| = 0.$$

Take an arbitrary solution $x(t)$ and write it in the following way:

$$\begin{aligned} x(t) &= \frac{x(t)}{\|x(t_0)\|} \cdot \frac{\|x(t_0)\|}{\Delta/2} \cdot \frac{\Delta}{2} \\ &= z(t) \frac{2\|x(t_0)\|}{\Delta}. \end{aligned}$$

Obviously,

$$\|z(t_0)\| = \Delta/2 < \Delta;$$

therefore,

$$\|z(t)\| \underset{t\to\infty}{\longrightarrow} 0,$$

and, thus,

$$\|x(t)\| \underset{t\to\infty}{\longrightarrow} 0.$$

Conversely, let

$$\|x(t)\| \underset{t\to\infty}{\longrightarrow} 0 \qquad \text{for any solution } x(t).$$

For each solution there exists a moment of time $T > t_0$ such that

$$\|x(t)\| < 1 \qquad \text{for} \qquad t \geqslant T.$$

On the interval $[t_0, T]$ the boundedness of $\|x(t)\|$ follows from continuity. According to Theorem 4.1.2, the trivial solution is stable. But for any

$$\|x(t_0)\| < \infty,$$

according to our condition, we have

$$\|x(t)\| \underset{t\to\infty}{\longrightarrow} 0,$$

i.e., $\Delta = \infty$ and the asymptotic stability is proved. □

COROLLARY 4.1.3. *An asymptotically stable linear system is globally asymptotically stable.*

REMARK 4.1.3. The rate at which solutions tend to zero is different for different linear systems. Indeed, consider the following examples:

$$\begin{aligned} &1. \quad \dot{x} = -x/t, \qquad && x(t) = c/t, \\ &2. \quad \dot{x} = -x, && x = ce^{-t}. \end{aligned}$$

The following version of asymptotic stability is frequently used in applications.

DEFINITION 4.1.5. A linear homogeneous system is said to be *exponentially stable* if there exist constants $d > 0$ and $\varepsilon > 0$ such that for any solution $x(t)$ we have

$$\|x(t)\| \leqslant d\|x(t_0)\|e^{-\varepsilon(t-t_0)}, \qquad t \geqslant t_0.$$

## §2. On stability of linear homogeneous sytems whose coefficients are constant, periodic, or satisfy the Lappo-Danilevskiĭ condition

As before, we consider a system

$$\dot{x} = A(t)x, \qquad x \in \mathbb{C}^n. \tag{4.2.1}$$

The theorems proven in the previous section allow us to obtain quite constructive stability criteria for systems indicated in the title.

THEOREM 4.2.1. *A linear homogeneous system with constant coefficients is*

1) *stable if and only if all the eigenvalues of the coefficient matrix have nonpositive real parts, and simple elementary divisors correspond to the eigenvalues with zero real part,*

2) *asymptotically stable if and only if all the eigenvalues of the coefficient matrix have negative real parts.*

The validity of this theorem easily follows from the analysis of the dependence of the behavior of solutions of a linear homogeneous autonomous system on the character of eigenvalues of the coefficient matrix carried out in Remark 1.3.1 and, further, from Theorems 4.1.2 and 4.1.3.

In spite of enticing simplicity of the formulation of Theorem 4.2.1, its application encounters difficulties of algebraic character: it requires finding the location of the roots of the characteristic equation

$$\operatorname{Det}(A - \lambda E) = 0 \tag{4.2.2}$$

on the complex plane.

As is known, for $n > 4$ this problem can be solved only approximately and even for $n = 3$, $n = 4$ approximations are often used. However, there is a number of results formulated in terms of the matrix $A$ that can be useful for us. We give some of them omitting the proofs, which can be found in the references indicated.

**The Hurwitz criterion [19, 16]**. We write equation (4.2.2) in the form

$$a_0\lambda^n + a_1\lambda^{n-1} + \cdots + a_{n-1}\lambda + a_n = 0, \tag{4.2.2'}$$

and assume that all the coefficients of the characteristic equation are real; moreover, $a_0 > 0$. Let us form an $n \times n$ square matrix

$$\begin{pmatrix} a_1 & a_3 & a_5 & \dots & 0 \\ a_0 & a_2 & a_4 & \dots & 0 \\ 0 & a_1 & a_3 & \dots & 0 \\ \dots & \dots & \dots & \dots & \dots \\ 0 & 0 & 0 & \dots & a_n \end{pmatrix}$$

with these coefficients as elements according to the following scheme. In the first row there are coefficients with odd indices beginning with $a_1$. The elements of each subsequent row are formed from the elements of the previous one by reducing the index by one, and, if necessary, adding on the right of the row the next coefficient of the polynomial (4.2.2′) so that each odd row contains all the coefficients with odd indices and each even row—with even ones. As a result of this construction, the coefficients

$$a_1, a_2, \dots, a_n$$

must be on the main diagonal, and all the elements of the last column except for the last one are zeros. Consider the principal minors of this matrix,

$$\Delta_1 = a_1, \Delta_2 = a_1a_2 - a_0a_3, \dots, \Delta_n = a_n\Delta_{n-1}. \tag{4.2.3}$$

THEOREM 4.2.2 (Hurwitz). *All the roots of equation (4.2.2′) with real coefficients for $a_0 > 0$ have negative real parts if and only if all the principal minors (4.2.3) are positive.*

COROLLARY 4.2.1. *The Viète formula straightforwardly implies the following statements:*

1) *if for $a_0 > 0$ all the roots of equation (4.2.2′) have negative real parts, then all the coefficients $a_1, a_2, \dots, a_n$ are positive,*

2) *if for $a_0 > 0$ at least one of the coefficients $a_1, \dots, a_n$ is negative, then equation (4.2.2′) has roots with positive real parts.*

*The following results can be found in* **[27]**:

3) *if a real matrix $A = \{a_{ij}\}$ is such that*

$$a_{ij} \geqslant 0 \quad \text{for all} \quad i, j = 1, \dots, n, \quad i \neq j,$$

*and there exist positive numbers $\alpha_1, \alpha_2, \dots, \alpha_n$ such that*

$$\sum_{j=1}^{n} \alpha_j a_{ij} < 0, \qquad i = 1, \dots, n, \tag{4.2.4}$$

*then all the eigenvalues of the matrix $A$ have negative real parts,*

4) *if the matrix $A$ is complex, then inequalities* (4.2.4) *should be replaced with the inequalities*

$$\sum_{j\neq i} \alpha_j |a_{ij}| < -\alpha_i \operatorname{Re} a_{ii}, \qquad i = 1, \dots, n,$$

5) *if the matrix $A$ is complex and is such that*

$$\operatorname{Re} a_{ii} < -\sum_{j\neq i} |a_{ij}|, \qquad i = 1, \dots, n, \tag{4.2.5}$$

*then all the eigenvalues of the matrix $A$ have negative real parts. If* $\operatorname{Im} A = 0$ *and inequalities* (4.2.5) *hold, then all the real parts of the eigenvalues of the matrix $A$ are negative if and only if* $\operatorname{Det} A \neq 0$.

Now we pass to the case of periodic coefficients in system (4.2.1).

THEOREM 4.2.3. *A linear homogeneous system with periodic coefficients is*

1) *stable if and only if all its multipliers belong to the closed unit disc, and simple elementary divisors correspond to the multipliers lying on the unit circle,*

2) *asymptotically stable if and only if all its multipliers belong to the interior of the unit disc.*

The validity of this statement follows from Remark 1.4.8 and Theorems 4.1.2, 4.1.3.

Note that the problem of location of multipliers on the complex plane is very complicated and there are no efficient general methods for solving it. Usually, one turns to methods of small parameter, or approximation methods for determining the monodromy matrix and then its eigenvalues.

Now we pass to systems known as *Lappo-Danilevskiĭ systems.* This name is attached to system (4.2.1) if its coefficient matrix commutes with its integral, i.e., the equality

$$A(t) \int_s^t A(\tau)\, d\tau = \int_s^t A(\tau)\, d\tau A(t) \tag{4.2.6}$$

is valid for all $t, s \in [t_0, \infty)$.

LEMMA 4.2.1. *A linear homogeneous system whose coefficient matrix satisfies the condition* (4.2.6) *has a fundamental matrix $X(t)$ of the form*

$$X(t) = \exp \int_{t_0}^t A(\tau)\, d\tau. \tag{4.2.7}$$

PROOF. We verify that the matrix (4.2.7) satisfies the equation

$$\dot{X} = A(t)X, \qquad X(t_0) = E. \tag{4.2.8}$$

By definition,

$$e^{\int_{t_0}^t A(\tau)\, d\tau} = E + \int_{t_0}^t A(\tau)\, d\tau + \dots + \frac{\left[\int_{t_0}^t A(\tau)\, d\tau\right]^k}{k!} + \cdots. \tag{4.2.9}$$

The series (4.2.9) converges for all $t \in [t_0, \infty)$, and for $t \in [t_0, T]$ it converges uniformly. Let us differentiate the series term by term, noting that

$$\frac{d}{dt}\frac{\left[\int_{t_0}^{t} A(\tau)\,d\tau\right]^k}{k!} = \frac{1}{k!}\Big(A(t)\underbrace{\int_{t_0}^{t} A\,d\tau\cdots\int_{t_0}^{t} A\,d\tau}_{k-1} + \int_{t_0}^{t} A\,d\tau A(t)\underbrace{\int_{t_0}^{t} A\,d\tau\cdots\int_{t_0}^{t} A\,d\tau}_{k-1}$$
$$+\cdots+\underbrace{\int_{t_0}^{t} A\,d\tau\cdots\int_{t_0}^{t} A\,d\tau}_{k-1} A(t)\Big).$$

This notation shows that it is impossible to obtain a fundamental matrix in the form (4.2.7) for an arbitrary linear system. However, in the case of (4.2.6) we have

$$\frac{d}{dt}\frac{\left[\int_{t_0}^{t} A(\tau)\,d\tau\right]^k}{k!} = \frac{1}{(k-1)!}A(t)\left[\int_{t_0}^{t} A(\tau)\,d\tau\right]^{k-1}.$$

Differentiating the series (4.2.9) term by term, we obtain the same series multiplied on the left by the matrix $A(t)$. Hence, the identity (4.2.8) is satisfied and $X(t)$ is a fundamental matrix. □

THEOREM 4.2.4. *Let the linear system* (4.2.1) *be such that*
1) *condition* (4.2.6) *is satisfied,*
2) *there exists*

$$\lim_{t\to\infty}\frac{1}{t}\int_{t_0}^{t} A(\tau)\,d\tau = \Lambda, \tag{4.2.10}$$

3) *all the eigenvalues of the matrix* $\Lambda$ *have negative real parts.*
*Then system* (4.2.1) *is asymptotically stable.*

PROOF. We show that the matrices $A(t)$ at different values of the argument commute. In order to do this, we differentiate the identity (4.2.6) with respect to $s$:

$$A(t)[\ \ A(s)] = -A(s)A(t).$$

Using this fact, we write

$$\int_{t_0}^{t} A(t_1)\,dt_1\frac{1}{s}\int_{t_0}^{s} A(t_2)\,dt_2 = \frac{1}{s}\int_{t_0}^{t} dt_1\int_{t_0}^{s} A(t_1)A(t_2)\,dt_2$$
$$= \frac{1}{s}\int_{t_0}^{t} dt_1\int_{t_0}^{s} A(t_2)A(t_1)\,dt_2$$
$$= \frac{1}{s}\int_{t_0}^{s} A(t_2)\,dt_2\int_{t_0}^{t} A(t_1)\,dt_1.$$

Letting $s$ tend to infinity, from the last identity we obtain

$$\int_{t_0}^{t} A(\tau)\,d\tau\Lambda = \Lambda\int_{t_0}^{t} A(\tau)\,d\tau. \tag{4.2.11}$$

From this, according to (4.2.10), we have

$$\frac{1}{t}\int_{t_0}^{t} A(\tau)\,d\tau = \Lambda + B(t), \qquad \text{and} \qquad \|B(t)\| \underset{t\to\infty}{\longrightarrow} 0.$$

We show that

$$\Lambda B(t) = B(t)\Lambda, \tag{4.2.12}$$

$$\Lambda B(t) = \Lambda\left[\frac{1}{t}\int_{t_0}^{t} A(\tau)\,d\tau - \Lambda\right] \overset{(4.2.11)}{=} \left[\frac{1}{t}\int_{t_0}^{t} A(\tau)\,d\tau - \Lambda\right]\Lambda = B(t)\Lambda.$$

For any solution $x(t)$ of system (4.2.1), by Lemma 4.2.1, we have

$$x(t) = e^{\int_{t_0}^{t} A(\tau)\,d\tau}x(t_0) = e^{\Lambda t + B(t)t}x(t_0) \overset{(4.2.12)}{=} e^{\Lambda t}e^{B(t)t}x(t_0).$$

Let the real parts of the eigenvalues of the matrix $\Lambda$ not exceed $\alpha < 0$. Take $\varepsilon > 0$ so that $\alpha + 2\varepsilon < 0$ and define $T > t_0$ from the condition $\|B(t)\| < \varepsilon$ for $t \geqslant T$. To estimate $\|x(t)\|$, we use the estimate

$$\|e^{B(t)t}\| \leqslant e^{\|B(t)\|t},$$

which easily follows from the series for the matrix exponential, as well as the estimate (1.3.10) of the matriciant of an autonomous system. Therefore, for $t \geqslant T$ we have

$$\begin{aligned}\|x(t)\| &\leqslant \|e^{\Lambda t}\| \cdot \|e^{B(t)t}\| \cdot \|x(t_0)\| \\ &\leqslant M_\varepsilon e^{(\alpha+\varepsilon)T} e^{\|B(t)\|t}\|x(t_0)\| \\ &\leqslant M_\varepsilon e^{(\alpha+2\varepsilon)t}.\end{aligned}$$

Hence,

$$\|x(t)\| \to 0, \qquad t \to \infty, \qquad \text{by virtue of} \qquad \alpha + 2\varepsilon < 0. \quad \square$$

## §3. Almost constant systems

In this section we consider three results concerning the influence of small variation of the coefficients of an autonomous system on its stability. All these theorems are particular cases of more general theorems (for variable coefficients), which will be proved in the next section. In the theorems of this section the conditions are, naturally, simpler, the proofs are based on the fact that the system is autonomous and represent typical arguments for problems of this kind. It is useful to get acquainted with them.

Together with a system

$$\dot{x} = Ax, \tag{4.3.1}$$

where $x \in \mathbb{C}^n$ and $A$ is a constant matrix, we shall consider the perturbed system

$$\dot{y} = [A + B(t)]y, \qquad B \in C[t_0, \infty), \tag{4.3.2}$$

where the matrix $B(t)$ is assumed to be small in some sense.

THEOREM 4.3.1. *Let system* (4.3.1) *be stable and let*

(4.3.3) $$\int_{t_0}^{t} \|B(\tau)\|\,d\tau < \infty.$$

*Then system* (4.3.2) *is also stable.*

PROOF. Let

$$X(t, t_0) = \exp A(t - t_0)$$

be the fundamental matrix of system (4.3.1). By the method of variation of parameters, for any solution $y(t)$ of system (4.3.2), we have

$$y(t) = X(t, t_0)y(t_0) + \int_{t_0}^{t} X(t, \tau)B(\tau)y(\tau)\,d\tau,$$

or

(4.3.4) $$\|y(t)\| \leqslant \|X(t, t_0)\| \cdot \|y(t_0)\| + \int_{t_0}^{t} \|X(t, \tau)\| \cdot \|B(\tau)\| \cdot \|y(\tau)\|\,d\tau.$$

Due to the stability of system (4.3.1), by Theorem 4.1.2 there exists a constant $K > 0$ such that

$$\|X(t, \tau)\| \leqslant K \qquad \text{for} \qquad t \geqslant \tau \geqslant t_0.$$

Thus,

$$\|y(t)\| \leqslant K\|y(t_0)\| + K \int_{t_0}^{t} \|B(\tau)\| \cdot \|y(\tau)\|\,d\tau.$$

By the Gronwall-Bellman lemma (see Appendix), we have

$$\|y(t)\| \leqslant K\|y(t_0)\| \exp K \int_{t_0}^{t} \|B(\tau)\|\,d\tau \overset{(4.3.3)}{<} \infty.$$

By Theorem 4.1.2, from the boundedness of the solutions of system (4.3.2) we obtain its stability. □

REMARK 4.3.1. In §4 of this chapter we shall consider Perron's Example 4.4.1 refuting the validity of Theorem 4.3.1 and of the next Theorem 4.3.2 in the case of a variable matrix $A(t)$.

THEOREM 4.3.2. *If system* (4.3.1) *is asymptotically stable and*

(4.3.5) $$\|B(t)\| \to 0, \qquad t \to \infty,$$

*then the perturbed system* (4.3.2) *is also asymptotically stable.*

PROOF. An autonomous linear system can be asymptotically stable if and only if all the eigenvalues of the coefficient matrix have negative real parts. Let $\alpha = \max_j \operatorname{Re} \lambda_j$, where $\lambda_j$, $j = 1, \dots, n$, are eigenvalues of the matrix $A$. Then (see (1.3.10)) the estimate

(4.3.6) $$\|e^{A(t-\tau)}\| \leqslant C_\varepsilon e^{(\alpha+\varepsilon)(t-\tau)}, \qquad t \geqslant \tau,$$

is valid. We use (4.3.6) in the inequality (4.3.4) to obtain the estimate

$$\|y(t)\| \leqslant C_\varepsilon e^{(\alpha+\varepsilon)(t-t_0)}\|y(t_0)\| + \int_{t_0}^{t} C_\varepsilon \|B(\tau)\| e^{(\alpha+\varepsilon)(t-\tau)}\|y(\tau)\|\,d\tau,$$

and choose an $\varepsilon > 0$ such that

$$\alpha + 2\varepsilon < 0. \tag{4.3.7}$$

Thus,

$$\|y(t)\|e^{-(\alpha+\varepsilon)t} \leqslant C_\varepsilon e^{-(\alpha+\varepsilon)t_0}\|y(t_0)\| + \int_{t_0}^{t} C_\varepsilon \|B(\tau)\|e^{-(\alpha+\varepsilon)\tau}\|y(\tau)\|\,d\tau$$

and, further, by the Gronwall-Bellman lemma (see Appendix), we have

$$\|y(t)\|e^{-(\alpha+\varepsilon)t} \leqslant C_\varepsilon\|y(t_0)\|e^{-(\alpha+\varepsilon)t_0}e^{\int_{t_0}^{t} C_\varepsilon\|B(\tau)\|\,d\tau}. \tag{4.3.8}$$

By the generalized L'Hospital rule **[34]** and the condition (4.3.5), we have

$$\lim_{t\to\infty}\frac{C_\varepsilon\int_{t_0}^{t}\|B(\tau)\|\,d\tau}{t-t_0} = \lim_{t\to\infty}\frac{C_\varepsilon\|B(t)\|}{1} = 0,$$

and for $t \geqslant T$ we have

$$C_\varepsilon\int_{t_0}^{t}\|B(\tau)\|\,d\tau < \varepsilon(t-t_0).$$

From (4.3.8) for $t \geqslant T$ we obtain

$$\|y(t)\|e^{-(\alpha+\varepsilon)t} \leqslant C_\varepsilon\|y(t_0)\|e^{-(\alpha+\varepsilon)t_0}e^{\varepsilon(t-t_0)},$$

and, finally,

$$\|y(t)\| \leqslant C_\varepsilon\|y(t_0)\|e^{(\alpha+2\varepsilon)(t-t_0)}.$$

This, by virtue of (4.3.7), implies our claim. □

An example shows that stability (not asymptotic) may be lost if

$$\|B(t)\| \underset{t\to\infty}{\longrightarrow} 0.$$

EXAMPLE 4.3.1. The system $\dot{x}_1 = 0$, $\dot{x}_2 = 0$ is stable since its solution is

$$x_1(t) = C_1, \qquad x_2(t) = C_2.$$

Consider the perturbation

$$B(t) = \begin{pmatrix} 0 & 0 \\ 1/t & 0 \end{pmatrix}.$$

The system $\dot{y}_1 = 0$, $\dot{y}_2 = y_1/t$ is unstable, since

$$y_1(t) = C_1, \qquad y_2(t) = C_1\ln t + C_2 \to \infty \qquad \text{as} \qquad t\to\infty, \quad C_1 \neq 0.$$

The following theorem follows from the one just proved, but is quite constructive.

THEOREM 4.3.3. *Let system* (4.3.1) *be asymptotically stable and let $\alpha < 0$ be the greatest real part of the eigenvalues of the matrix $A$; then, under the condition*

$$\|B(t)\| \leqslant C_1 < (-\alpha-\varepsilon)/C_\varepsilon, \qquad t \geqslant t_0, \tag{4.3.9}$$

*the perturbed system* (4.3.2) *is also asymptotically stable.*

*Here $\varepsilon > 0$ is such that $\alpha + \varepsilon < 0$ and $C_\varepsilon$ is taken from the estimate* (1.3.10).

REMARK 4.3.2. If the eigenvalues of the matrix $A$ are simple, then the condition (4.3.9) is changed to the condition

$$\|B(t)\| \leqslant C_1 < -\alpha/M, \qquad t \geqslant t_0, \tag{4.3.10}$$

which follows from the estimate (1.3.12).

PROOF. We repeat the proof of the previous theorem, omitting the requirement (4.3.7), up to and including the relation (4.3.8). From (4.3.8), we have

$$\|y(t)\|e^{-(\alpha+\varepsilon)t} \leqslant C_\varepsilon\|y(t_0)\|e^{-(\alpha+\varepsilon)t_0}e^{C_\varepsilon C_1(t-t_0)},$$

or

$$\|y(t)\| \leqslant C_\varepsilon\|y(t_0)\|e^{(\alpha+\varepsilon+C_1C_\varepsilon)(t-t_0)}.$$

By virtue of the estimate (4.3.9),

$$\alpha+\varepsilon+C_1C_\varepsilon < \alpha+\varepsilon-\alpha-\varepsilon = 0.$$

Thus,

$$\|y(t)\| \to 0, \qquad t \to \infty. \qquad \square$$

We illustrate Theorems 4.3.2 and 4.3.3.

EXAMPLE 4.3.2. 1. Consider the system

$$\dot{x}_1 = -3x_1 + 2x_2,$$
$$\dot{x}_2 = x_1 - 4x_2;$$

we have

$$X(t,0) = e^{At} = \frac{1}{3}\begin{pmatrix} e^{-5t}+2e^{-2t} & -2e^{-5t}+2e^{-2t} \\ -e^{-5t}+e^{-2t} & 2e^{-5t}+e^{-2t} \end{pmatrix},$$

$$\|e^{At}\| \leqslant \frac{5}{3}e^{-2t}, \qquad t \geqslant 0.$$

Under what perturbations $B(t)$ does the system $\dot{x} = [A + B(t)]x$ preserve the asymptotic stability? The answer is provided by the following conditions:

a) $\|B(t)\| \to 0$, $t \to \infty$ (Theorem 4.3.2),

b) $\|B(t)\| \leqslant C_1 < 2 \cdot 3/5 = 6/5$, $t \geqslant 0$ (Theorem 4.3.3, condition (4.3.10)).

2. Consider the system

$$\dot{x}_1 = -x_1 + x_2,$$
$$\dot{x}_2 = -x_1 - 3x_2;$$

we have

$$X(t,0) = e^{At} = \begin{pmatrix} (1+t)e^{-2t} & te^{-2t} \\ -te^{-2t} & (1-t)e^{-2t} \end{pmatrix},$$

$$\|e^{At}\| \leqslant (1+2t)e^{-2t}, \qquad t \geqslant 0.$$

Take $\varepsilon > 0$ and obtain the estimate (1.3.10):

$$\|e^{At}\| \leqslant (1+2t)e^{-\varepsilon t}e^{-2t}e^{\varepsilon t} \leqslant \frac{2}{\varepsilon}e^{(\varepsilon-2)/2}e^{(-2+\varepsilon)t} = C_\varepsilon e^{(-2+\varepsilon)t}.$$

Here

$$C_\varepsilon = \max_{t\in\mathbb{R}_+}(1+2t)e^{-\varepsilon t} = \frac{2}{\varepsilon}e^{(\varepsilon-2)/2}.$$

The system $\dot{x} = [A + B(t)]x$ remains asymptotically stable if

a) $\|B(t)\| \to 0$, $t \to \infty$ (Theorem 4.3.2),

b) $\|B(t)\| \leqslant C_1 < \dfrac{\varepsilon(2-\varepsilon)e}{2e^{\varepsilon/2}}$ (Theorem 4.3.3).

Theorem 4.3.3 gives an explicit bound for the perturbations under which the characteristic exponents of the perturbed system remain negative. However, if $\alpha$ is close to zero, then this theorem sharply loses its practical value, and, if $\alpha = 0$, the theorem is inapplicable at all. At the same time, small perturbations of the coefficients of the system may diminish the characteristic exponents and even make them negative. One of the approaches to problems of this sort is the method of small parameter. We show its practical application in our case by presenting a result of Malkin [**26**]. A more general approach can be found in [**21**].

We write system (4.3.2) in the form

$$\dot{y} = [A + \mu B(t)]y, \qquad B \in C[0, \infty), \tag{4.3.11}$$

and assume from the outset that

$$A = \operatorname{diag}[\alpha_1, \dots, \alpha_n], \qquad \alpha_i \neq \alpha_j, \quad i \neq j,$$

and that some of its eigenvalues have zero real parts, but there are no eigenvalues with positive ones; $\mu$ is a small parameter,

$$\sup_{t \geqslant 0} \|B(t)\| < \infty.$$

The idea of the method is to increase the order of smallness with respect to $\mu$ of the variable terms in the coefficients of the system. The method is carried out in the following way: by means of the Lyapunov transformation

$$z = [E + \mu V(t)]y, \tag{4.3.12}$$

system (4.3.11) is reduced to the system

$$\dot{z} = [A + \mu C + \mu^2 \Psi(t)]z. \tag{4.3.13}$$

Here $\Psi(t)$ is a matrix bounded on $\mathbb{R}_+$; $A + \mu C$, where $C$ is chosen so that $V(t)$ is bounded, plays the role of the constant matrix. The advantage of the new system is that the maximal real part of the eigenvalues of the matrix $A + \mu C$ may turn out to be negative, its order being no higher than $\mu$, and the order of perturbation is $\mu^2$; then Theorem 4.3.3 is applicable. Note that the condition that $A$ be diagonal is assumed to simplify calulations. The case when $\alpha = \max_i \operatorname{Re} \lambda_i$ is sufficiently small can be reduced to ours by including it in the term of order $\mu$.

THEOREM 4.3.4. *If the matrix $B(t)$ is such that*

1) *there exists a constant $c_{ss}$ such that*

$$c_{ss} t - \int_0^t b_{ss}(\tau)\, d\tau < \infty, \qquad s = 1, \dots, n, \tag{4.3.14}$$

2) *in the case when*

$$\lambda_s - \lambda_k = i\beta_{sk}, \qquad s, k = 1, \dots, n,$$

*the following integrals are bounded:*

(4.3.15)
$$\int_0^t b_{sk}(\tau)\cos\beta_{sk}\tau\,d\tau < \infty,$$
$$\int_0^t b_{sk}(\tau)\sin\beta_{sk}\tau\,d\tau < \infty,$$

*then there exists a Lyapunov transformation reducing system* (4.3.11) *to system* (4.3.13).

REMARK 4.3.3. The condition (4.3.14) is *a fortiori* satisfied for periodic and quasiperiodic functions. The conditions (4.3.15) are satisfied for these functions if their expansions do not have resonant harmonics $\cos\beta_{sk}t$ and $\sin\beta_{sk}t$.

PROOF. Substituting the change of variable (4.3.12) in (4.3.13) and taking into account (4.3.11), we obtain

$$(E+\mu V)(A+\mu B)y+\mu\dot V y=(A+\mu C+\mu^2\Psi)(E+\mu V)y.$$

Equating the terms containing $\mu$ to the first power, we obtain

$$VA+B+\dot V=C+AV,$$

or

(4.3.16)
$$\dot V=AV-VA+C-B.$$

We show that if the conditions of the theorem are satisfied and the matrix $C$ is chosen appropriately, then the matrix equation (4.3.16) has a solution bounded for $t>0$. By virtue of (4.3.16), the fact that the matrix $V(t)$ is bounded implies that the matrix $\dot V(t)$ is also bounded; the boundedness of $[E+\mu V(t)]^{-1}$ follows from the fact that $\operatorname{Det}(E+\mu V(t))$ is bounded away from zero for sufficiently small $\mu$. Equation (4.3.16) splits into $n^2$ scalar linear equations defining $v_{sk}(t)$, i.e., the elements of the matrix $V(t)$:

(4.3.17)
$$\dot v_{sk}=(\lambda_s-\lambda_k)v_{sk}+c_{sk}-b_{sk}(t),\qquad s,k=1,\dots,n.$$

Our goal is to find solutions of these equations bounded for $t\geqslant 0$; consider the following cases:

1. $s=k$, $\dot v_{sk}=c_{ss}-b_{ss}(t)$; therefore

$$v_{ss}(t)=c_{ss}t-\int_0^t b_{ss}(\tau)\,d\tau<\infty,\qquad \text{by (4.3.14)}.$$

2. $s\neq k$. Set

$$c_{sk}=0,\qquad \lambda_s-\lambda_k=\alpha_{sk}+i\beta_{sk}.$$

As $v_{sk}(t)$, take

$$v_{sk}(t)=-e^{(\alpha_{sk}+i\beta_{sk})t}\int_a^t e^{-(\alpha_{sk}+i\beta_{sk})\tau}b_{sk}(\tau)\,d\tau,$$

where

$$a=\begin{cases}0, & \text{if}\quad \alpha_{sk}\leqslant 0,\\ \infty, & \text{if}\quad \alpha_{sk}>0.\end{cases}$$

Let us verify that the function $v_{sk}(t)$ for $\alpha_{sk}\neq 0$ is bounded for $t\geqslant 0$. We assume that

$$|b_{sk}(t)|\leqslant M,\qquad t\geqslant 0.$$

a) In the case $\alpha_{sk} < 0$ we have the estimate

$$|v_{sk}(t)| \leqslant Me^{\alpha_{sk}t} \int_0^t e^{-\alpha_{sk}\tau}\, d\tau = \frac{M}{-\alpha_{sk}} e^{\alpha_{sk}t}(e^{-\alpha_{sk}t} - 1) \leqslant \frac{M}{|\alpha_{sk}|}.$$

b) For $\alpha_{sk} > 0$ we obtain

$$|v_{sk}(t)| \leqslant Me^{\alpha_{sk}t} \int_t^\infty e^{-\alpha_{sk}\tau}\, d\tau = \frac{M}{\alpha_{sk}}.$$

c) For $\alpha_{sk} = 0$, $\beta_{sk} \neq 0$ we have

$$v_{sk}(t) = e^{i\beta_{sk}t} \int_0^t (\cos\beta_{sk}\tau - i\sin\beta_{sk}\tau) b_{sk}(\tau)\, d\tau.$$

Now the boundedness follows from the conditions (4.3.15). □

REMARK 4.3.4. If the matrix $\Psi(t)$ possesses the properties (4.3.14) and (4.3.15), then we can subject the system (4.3.13) to a transformation of the form (4.3.12) and obtain that the variable terms are of order $\mu^3$. If we can carry out these arguments $k$ times, then system (4.3.11) will pass into the system

$$\dot{z} = Az + \sum_{l=1}^{k} \mu^l C_l z + \mu^{k+1}\widetilde{\Psi}(t)z.$$

As an illustration, consider the following example.

EXAMPLE 4.3.3 (Malkin [26]).

$$\dot{y}_1 = \mu(-1 + 2\sin t)y_1 + \mu y_2,$$
$$\dot{y}_2 = -y_2 + \mu y_1.$$

Here

$$A = \begin{pmatrix} 0 & 0 \\ 0 & -1 \end{pmatrix}, \qquad \lambda_1 = 0, \quad \lambda_2 = -1, \qquad B(t) = \begin{pmatrix} -1 + 2\sin t & 1 \\ 1 & 0 \end{pmatrix}.$$

For the elements of the matrix $V(t)$ from equations (4.3.17) we have

$$\dot{v}_{11} = c_{11} + 1 - 2\sin t,$$
$$\dot{v}_{12} = v_{12} - 1 + c_{12},$$
$$\dot{v}_{22} = c_{22},$$
$$\dot{v}_{21} = -v_{21} - 1 + c_{21}.$$

Setting $c_{12} = 0$, $c_{21} = 0$, $c_{11} = -1$, $c_{22} = 0$, we take the following solutions:

$$v_{11} = 2\cos t, \qquad v_{22} = 0, \qquad v_{12} = 1, \qquad v_{21} = -1.$$

Thus, the transformation (4.3.12) is defined, and the initial system reduces to the form

$$\dot{z}_1 = -\mu z_1 + \mu^2(\psi_{11}z_1 + \psi_{12}z_2),$$
$$\dot{z}_2 = -z_2 + \mu^2(\psi_{21}z_1 + \psi_{22}z_2);$$

its autonomous part has the eigenvalues $-1$, $-\mu$, and the variable part is of order $\mu^2$.

Here

$$\psi_{11} = [1 + 2\sin 2t + \mu(1 + 2\cos t)]/\Delta,$$
$$\psi_{12} = [1 + 2\cos t + \mu(-1 + 4\cos^2 t - \sin 2t + 2\cos t)]/\Delta,$$
$$\psi_{21} = (1 - 2\sin t - \mu)/\Delta,$$
$$\psi_{22} = [-1 + \mu(-1 + 2\sin t - 2\cos t)]/\Delta,$$
$$\Delta = 1 + 2\mu\cos t + \mu^2.$$

By Theorem 4.3.3 and Remark 4.3.4, the characteristic exponents of this system are negative under the condition

$$\mu^2\|\psi(t)\| < \mu;$$

Malkin noted that this condition is satisfied at least for $\mu < 1/9$. We note that this example has become some sort of indicator of the effectivity of the estimates of the characteristic exponents via the coefficients of the system.

In this chapter Malkin's example will be used two more times: to illustrate Lozinskiĭ's method (Example 4.6.4) and Yakubovich's method (Example 4.6.6). In both cases, the domain of values of $\mu$ that ensure asymptotic stability is extended beyond $1/9$ to the right.

## §4. Uniformly stable and uniformly asymptotically stable linear systems

Consider a system

$$\dot{x} = A(t)x, \qquad x \in \mathbb{C}^n, \qquad A \in C(\mathbb{R}_+). \tag{4.4.1}$$

We begin with an example that shows that the results of the previous section obtained for constant coefficients do not automatically extend to variable coefficients.

EXAMPLE 4.4.1 (Perron [**4, 38**]). Consider the system

$$\begin{aligned} \dot{x}_1 &= -ax_1, \\ x_2 &= (\sin\ln t + \cos\ln t - 2a)x_2, \end{aligned} \qquad t \geqslant 1;$$

we have

$$x_1 = c_1 e^{-at},$$
$$x_2 = c_2 e^{t\sin\ln t - 2at}.$$

For

$$1 < 2a < 1 + \exp(-\pi)/2$$

the system is asymptotically stable. As a perturbed system, we consider the following one:

$$\dot{y}_1 = -ay_1,$$
$$\dot{y}_2 = -e^{-at}y_1 + (\sin\ln t + \cos\ln t - 2a)y_2;$$

we have

$$y_1 = c_1 e^{-at},$$
$$y_2 = e^{t\sin\ln t - 2at}\left[c_2 + c_1\int_1^t e^{-t_1\sin\ln t_1}\,dt_1\right].$$

Let us show that

$$\overline{\lim_{t\to\infty}} |y_2(t)| = \infty \qquad \text{if} \qquad c_1 \neq 0.$$

Consider the sequence

$$t_n = \exp(2n + 1/2)\pi, \qquad n = 1, 2, \dots,$$

and verify that it realizes the limit superior. Note that under the condition

$$\exp(2n - 1/2)\pi \leqslant t_1 \leqslant \exp(2n - 1/6)\pi$$

or, in other words, under the condition

$$t_n \exp(-\pi) \leqslant t_1 \leqslant t_n \exp(-2\pi/3)$$

we have

$$-\sin \ln t_1 \geqslant 1/2.$$

Therefore,

$$\int_1^{t_n} e^{-t_1 \sin \ln t_1}\, dt_1 > \int_{t_n e^{-\pi}}^{t_n e^{e^{-2\pi/3}}} e^{-t_1 \sin \ln t_1}\, dt_1 > (e^{-2\pi/3} - e^{-\pi}) t_n e^{(1/2) t_n e^{-\pi}}.$$

Hence,

$$|y_2(t)| > |c_1| e^{[1-2a+(1/2)\exp(-\pi)]t_n} t_n \underset{n\to\infty}{\longrightarrow} \infty.$$

Let us discuss the example. The initial system is asymptotically stable, the perturbation satisfies the condition of Theorems 4.3.1 and 4.3.2, but neither asymptotic nor standard stability have been preserved under this perturbation. The reason is that here the stability in the standard sense is not sufficient; it is necessary that this property be uniform with respect to the initial moment $t_0$. In the case of constant coefficients the stability is automatically uniform; this allowed us to obtain the results of §3.

In §1 of this chapter we gave definitions from stability theory, considered the unperturbed solution $x(t, t_0, x_0)$, and the perturbed one $x(t, t_0, \eta)$; $t_0$ was fixed and the value of $\delta$ (see Definition 4.1.1) actually depended not only on $\varepsilon$, but also on $t_0$; the stability itself did not depend on $t_0$. The definition given below differs from the indicated ones in that $\delta$ depends only on $\varepsilon$, i.e., it is the same for any $t_0 \in \mathbb{R}_+$ (we use the notation from §1).

DEFINITION 4.4.1. A solution

$$x(t, t_0, x_0) \qquad (t_0 \in \mathbb{R}_+, \quad x_0 \in \mathbb{C}^n)$$

is said to be *uniformly stable* if for any $\varepsilon > 0$ there exists a $\delta > 0$, independent of the choice of $t_0$, such that the condition $\|x_0 - \eta\| < \delta$ implies the inequality

$$\|x(t, t_0, x_0) - x(t, t_0, \eta)\| < \varepsilon, \qquad t \geqslant t_0 \geqslant 0.$$

DEFINITION 4.4.2. A solution $x(t, t_0, x_0)$ is said to be *uniformly asymptotically stable* if

1) it is uniformly stable,
2) there exists a $\delta_0 > 0$ and for every $\varepsilon > 0$ there is a $T = T(\varepsilon) > 0$ such that the condition $\|x_0 - \eta\| < \delta_0$, $t_0 \in \mathbb{R}_+$ is arbitrary, implies the inequality

$$\|x(t, t_0, x_0) - x(t, t_0, \eta)\| < \varepsilon \qquad \text{for all} \qquad t \geqslant t_0 + T.$$

By means of arguments analogous to those in §1, it can be shown that the problem of uniform stability or uniform asymptotic stability of any solution of system (4.4.1) reduces to the analogous problem for the trivial solution $x \equiv 0$ of this system; this implies that all its solutions either do or do not possess simultaneously the properties mentioned above. This fact explains the terms *uniform stable linear system* or *uniform asymptotically stable linear system*.

We give examples of scalar equations to clarify these definitions.

EXAMPLE 4.4.2.

$$\dot{x} = -x, \qquad t \geqslant 0, \qquad x(t, t_0, x_0) = x_0 e^{-(t-t_0)}.$$

All the solutions tend to zero as $t \to \infty$; consequently, the equation is asymptotically stable.

Let us check Definition 4.4.1. Here

$$|x(t, t_0, x_0) - x(t, t_0, \eta)| = |x_0 - \eta| e^{-(t-t_0)}.$$

Obviously, the value of $\delta = \varepsilon$ does not depend on the specific value of $t_0$. Thus, we have uniform stability. We show that the second condition of Definition 4.4.2 is fulfilled. We have

$$|x(t, t_0, x_0) - x(t, t_0, \eta)| = e^{-(t-t_0)} |x_0 - \eta|.$$

Let us take a fixed $\delta_0$ and define

$$T(\varepsilon) = -\ln(\varepsilon/\delta_0) \qquad \text{for each} \qquad \varepsilon > 0;$$

then, under the condition that

$$|x_0 - \eta| < \delta_0$$

we have

$$|x(t, t_0, x_0) - x(t, t_0, \eta)| < e^{-(t-t_0)}\delta_0 \leqslant e^{\ln(\varepsilon/\delta_0)}\delta_0 = \varepsilon \qquad \text{for} \qquad t \geqslant t_0 + T(\varepsilon).$$

REMARK 4.4.1. In what follows, it will be shown that for autonomous systems, the stability and asymptotic stability in the standard sense are automatically uniform.

EXAMPLE 4.4.3. $\dot{x} = -x/t$, $t \geqslant 1$. In this equation

$$x(t, t_0, x_0) = x_0 t_0 / t,$$

the solution tends to zero as $t \to \infty$; consequently, the equation is asymptotically stable.

Let us check Definition 4.4.1. We have

$$|x(t, t_0, x_0) - x(t, t_0, \eta)| = |x_0 - \eta| t_0 / t.$$

Obviously, $\delta = \varepsilon$ does not depend on the choice of $t_0$, i.e., the equation is uniformly stable. Now we check whether the second condition of Definition 4.4.2 is satisfied. We have

$$|x(t, t_0, x_0) - x(t, t_0, \eta)| = \frac{t_0}{t} |x_0 - \eta| < \frac{t_0}{t} \delta_0.$$

Obviously, we cannot choose $T(\varepsilon)$ independently of $t_0$. Therefore, there is no uniform asymptotic stability.

EXAMPLE 4.4.4.

$$\dot{x} = (\sin \ln t + \cos \ln t - a)x, \qquad t \geqslant 1.$$

Here

$$x(t, t_0, x_0) = x_0 e^{t \sin \ln t - t_0 \sin \ln t_0 - a(t - t_0)}.$$

The equation is asymptotically stable for $a > 1$, since all the solutions $x(t) \to 0$ as $t \to \infty$. However, the equation is not uniformly stable. Indeed,

$$|x(t, t_0, x_0) - x(t, t_0, \eta)| = e^{t \sin \ln t - t_0 \sin \ln t_0 - a(t - t_0)} |x_0 - \eta|,$$

and we cannot indicate a $\delta > 0$ which would treat all $t_0 \geqslant 1$. Consider the sequence

$$t_n = \exp(2n + 1/2)\pi \qquad \text{for } t$$

and the sequence

$$t_0^{(n)} = \exp(2n - 1/2)\pi \qquad \text{for } t_0.$$

Note that

$$t_n - t_0^{(n)} = \exp 2n\pi(\exp(\pi/2) - \exp(-\pi/2)) \underset{n \to \infty}{\longrightarrow} \infty.$$

Hence,

$$\begin{aligned} &|x(t_n, t_0^{(n)}, x_0) - x(t_n, t_0^{(n)}, \eta)| \\ &\quad = \exp\{\exp(2n\pi + \pi/2) + \exp(2n\pi - \pi/2) \\ &\qquad - a[\exp(2n + 1/2)\pi - \exp(2n - 1/2)\pi]\}|x_0 - \eta| \\ &\quad = \exp\{\exp[2n\pi + \pi/2](1 - a + e^{-\pi} + ae^{-\pi})\}|x_0 - \eta| \underset{n \to \infty}{\longrightarrow} \infty, \end{aligned}$$

if

$$1 < a < (1 + e^{-\pi})(1 - e^{-\pi})^{-1}.$$

Let us formulate criteria of uniform stability and of uniform asymptotic stability in terms of the fundamental matrix $X(t)$ of system (4.4.1).

THEOREM 4.4.1. *System (4.4.1) is uniformly stable if and only if there exists a constant $D > 0$ such that for all $0 \leqslant s \leqslant t \leqslant \infty$ the inequality*

$$\|X(t, s)\| \equiv \|X(t)X^{-1}(s)\| \leqslant D \tag{4.4.2}$$

*holds.*

PROOF. It is sufficient to verify the statement of the theorem for the trivial solution $x \equiv 0$. We show that (4.4.2) implies its uniform stability. Any solution $x(t)$ can be written as

$$x(t) = X(t, t_0)x(t_0);$$

therefore, by (4.4.2), we obtain

$$\|x(t)\| \leqslant D\|x(t_0)\|.$$

For any $\varepsilon > 0$ we choose $\delta = \varepsilon/D$, and, independently of the specific value of $t_0 \in \mathbb{R}_+$, we have

$$\|x(t)\| < D\delta = D(\varepsilon/D) = \varepsilon, \qquad t \geqslant t_0.$$

Now let the solution $x \equiv 0$ be uniformly stable. Let us take $\varepsilon > 0$, choose $\delta > 0$ independently of $t_0$ according to it, and consider the solution $x(t)$ for any $t_0 \in \mathbb{R}_+$ such that

$$x(t_0) = (\delta/2, 0, \dots, 0)^\top,$$

i.e.,

$$x_1(t_0) = \delta/2, \qquad x_i(t_0) = 0, \qquad i \geqslant 2.$$

At the same time,

$$x(t) = X(t, t_0)x(t_0),$$

and, by virtue of the uniform stability, we have

$$\|X(t, t_0)x(t_0)\| < \varepsilon, \qquad t \geqslant t_0 \geqslant 0. \tag{4.4.3}$$

On the left-hand side of the inequality (4.4.3) under the sign of the norm there is the first column of the matrix $X(t, t_0)$ multiplied by $\delta/2$; this implies that the norm of this column does not exceed the value $2\varepsilon/\delta$. If the initial vector $x(t)$ is chosen so that

$$x_i(t_0) = \delta/2, \qquad x_j(t_0) = 0, \qquad i \neq j,$$

then we obtain a similar estimate for the $i$th column of the matrix $X(t, t_0)$; this shows that (4.4.2) is valid. □

COROLLARY 4.4.1. *For a linear system with constant coefficients the notions of stability and uniform stability are equivalent.*

Indeed, the stability of a linear system with constant coefficients, which form the matrix $A$, implies that the fundamental matrix $e^{At}$ is bounded (Theorem 4.1.2), i.e.,

$$\|e^{At}\| \leqslant D \qquad \text{for} \qquad t \geqslant 0.$$

For this system we have $X(t, s) = e^{A(t-s)}$. Therefore,

$$\|X(t, s)\| \leqslant D \qquad \text{for} \qquad t \geqslant s.$$

THEOREM 4.4.2. *System* (4.4.1) *is uniformly asymptotically stable if and only if there exist constants* $K > 0$, $\alpha > 0$ *independent of* $s$ *such that*

$$\|X(t, s)\| \leqslant Ke^{-\alpha(t-s)}, \qquad 0 \leqslant s \leqslant t < \infty. \tag{4.4.4}$$

PROOF. *Sufficiency.* Let (4.4.4) hold. We show that the trivial solution of system (4.4.1) is uniformly asymptotically stable. For any solution

$$x(t) = X(t, t_0)x(t_0)$$

of this system, by virtue of inequality (4.4.4), the following estimate holds:

$$\|x(t)\| \leqslant \|X(t, t_0)\|\|x(t_0)\| \leqslant Ke^{-\alpha(t-t_0)}\|x(t_0)\|, \qquad t \geqslant t_0 \geqslant 0. \tag{4.4.5}$$

Let us verify that the conditions of Definition 4.4.2 are satisfied. Uniform stability follows from the estimate

$$\|X(t, s)\| \leqslant K, \qquad t \geqslant s \geqslant 0.$$

We pass to the second condition. Let us take $\delta_0 = 1/K$ and define

$$T(\varepsilon) = (-1/\alpha)\ln \varepsilon$$

according to $\varepsilon > 0$ (for $\varepsilon > 1$ we take

$$T = (-1/\alpha)\ln(\varepsilon - [\varepsilon]));$$

then for $t \geqslant t_0 + T(\varepsilon)$ from (4.4.5) we have $\|x(t)\| < \varepsilon$.

*Necessity.* Let the trivial solution of system (4.4.1) be uniformly asymptotically stable. This means that there exists a $\delta_0$, for any $\varepsilon > 0$ there is a corresponding $T(\varepsilon)$, and for any solution $x(t)$ such that $\|x(s)\| < \delta_0$ at some moment $s \in \mathbb{R}_+$ we have

$$\|x(t)\| = \|X(t,s)x(s)\| < \varepsilon, \qquad t \geqslant s + T(\varepsilon). \tag{4.4.6}$$

By means of arguments similar to those in the proof of Theorem 4.4.1, from inequality (4.4.6) we obtain that

$$\|X(t,s)\| \to \infty \qquad \text{as} \qquad t - s \to \infty.$$

Let us determine the character of decrease more precisely. Fix $T > 0$ such that

$$\|X(s+T,s)\| \leqslant \theta < 1, \qquad s \in \mathbb{R}_+. \tag{4.4.7}$$

By virtue of the uniform stability, there exists a constant $M \geqslant 1$ such that

$$\|X(s+\tau,s)\| \leqslant M, \qquad s \in \mathbb{R}, \quad \tau \in [0,T]. \tag{4.4.8}$$

We pass to the proof of inequality (4.4.4) for

$$\|X(t,s)\|, \qquad t \geqslant s \geqslant 0.$$

Let

$$t = s + nT + \tau, \qquad n \in \mathbb{Z}_+, \quad \tau \in [0,T].$$

By the multiplicative property of the Cauchy matrix (see Chapter I, §2), we have

$$X(t,s) \equiv X(s+nT+\tau,s) = \prod_{k=n}^{1} X(s+kT+\tau, s+(k-1)T+\tau)\cdot X(s+\tau,s).$$

Therefore,

$$\|X(t,s)\| \leqslant \prod_{k=1}^{n} \|X(s+kT+\tau, s+(k-1)T+\tau)\|\cdot\|X(s+\tau,s)\|.$$

Taking logarithms in this inequality and using (4.4.7) and (4.4.8), we obtain

$$\begin{aligned}
\ln\|X(t,s)\| &\leqslant n\ln\theta + \ln M \\
&= \frac{1}{T}(t-s-\tau)\ln\theta + \ln M \\
&= \frac{1}{T}\ln\theta(t-s) - \frac{\tau}{T}\ln\theta + \ln M \\
&\leqslant \frac{1}{T}\ln\theta(t-s) - \ln\theta + \ln M.
\end{aligned}$$

Denoting

$$\alpha = -\frac{1}{T}\ln\theta, \qquad \ln K = \ln M - \ln\theta,$$

we have

$$\|X(t,s)\|Ke^{-\alpha(t-s)}, \qquad t \geqslant s \geqslant 0. \quad \square$$

COROLLARY 4.4.2. *For a linear system with constant coefficients the notion of asymptotic stability and uniform asymptotic stability are equivalent.*

Indeed, asymptotic stability of a linear system with constant coefficients means that all the eigenvalues $\lambda_i$, $i = 1, \dots, n$, of the matrix $A$ have negative real parts. Using the estimate (1.3.10), we write

$$\|X(t,s)\| = \|e^{A(t-s)}\| \leqslant K_\varepsilon e^{(\alpha+\varepsilon)(t-s)}, \qquad t \geqslant s \geqslant 0,$$

where

$$\max_i \operatorname{Re} \lambda_i = \alpha < 0, \qquad \alpha + \varepsilon < 0.$$

THEOREM 4.4.3. *If system* (4.4.1) *is stable* (*asymptotically stable*) *and reducible to a system with constant coefficients, then it is uniformly stable* (*uniformly asymptotically stable*).

PROOF. Let a Lyapunov transformation $x = L(t)y$ reduce system (4.4.1) to the autonomous system

$$\dot{y} = (L^{-1}(t)A(t)L(t) - L^{-1}(t)\dot{L}(t)) \equiv By. \tag{4.4.9}$$

The stability of the initial system implies that its fundamental matrix $X(t)$ is bounded; therefore, the fundamental matrix

$$Y(t) = L^{-1}(t)X(t)$$

of system (4.4.9) is also bounded; hence, this system is stable. By Corollary 4.4.1, system (4.4.9) is uniformly stable, i.e., there exists $D > 0$ such that we have

$$\|Y(t,s)\| \leqslant D \qquad \text{for all} \qquad t \geqslant s \geqslant 0.$$

Returning to the fundamental matrix of system (4.4.1), we obtain

$$\|X(t,s)\| = \|X(t)X^{-1}(s)\| = \|L(t)Y(t)Y^{-1}(s)L^{-1}(s)\| \leqslant C^2 D,$$

where

$$\|L(t)\| \leqslant C, \qquad \|L^{-1}(t)\| \leqslant C \qquad \text{for} \qquad t \subset \mathbb{R}_+.$$

Therefore, system (4.4.1) is uniformly stable. The statement on the uniform asymptotic stability is proved similarly by using Corollary 4.4.2 and the estimate

$$\|Y(t,s)\| \leqslant Ke^{-\alpha(t-s)}, \qquad t \geqslant s \geqslant 0, \quad K > 0, \quad \alpha > 0. \quad \square$$

COROLLARY 4.4.3. *If a system with periodic coefficients is stable* (*asymptotically stable*), *then it is uniformly stable* (*uniformly asymptotically stable*).

Together with system (4.4.1), we consider the perturbed system

$$\dot{y} = (A(t) + B(t))y, \qquad B \in C(\mathbb{R}_+), \tag{4.4.10}$$

and formulate conditions for the matrix $B(t)$ under which uniform stability and asymptotic stability are preserved. The first property does not change under absolutely integrable perturbations. Let us prove an analog of Theorem 4.3.1.

THEOREM 4.4.4. *Let*
1) *system* (4.4.1) *be uniformly stable,*
2) $\int_0^\infty \|B(\tau)\|\, d\tau \leqslant \beta < \infty$.
*Then system* (4.4.10) *is also uniformly stable.*

PROOF. Considering $B(t)y$ as a given function, we write the solution $y(t)$ of system (4.4.10) in the Cauchy form,

$$y(t) = X(t, t_0)y(t_0) + \int_{t_0}^{t} X(t, \tau)B(\tau)y(\tau)\, d\tau, \qquad t \geqslant t_0 \geqslant 0.$$

By the first condition of the theorem, for $X(t_1, t_0)$ we have the estimate (4.4.2). Therefore,

$$\|y(t)\| \leqslant D\|y(t_0)\| + D\int_{t_0}^{t} \|B(\tau)\|\|y(\tau)\|\, d\tau,$$

and, by the Gronwall-Bellman lemma (see Appendix), we have

$$\|y(t)\| \leqslant D\|y(t_0)\|e^{D\int_{t_0}^{t}\|B(\tau)\|\, d\tau} \leqslant D\|y(t_0)\|e^{D\beta}.$$

Now we verify that the trivial solution of system (4.4.10) is uniformly stable; this implies the statement of the theorem. We turn to Definition 4.4.1. We choose

$$\delta = \varepsilon/(D\exp(D\beta))$$

according to $\varepsilon > 0$, and from the last inequality it can be seen that if

$$\|y(t_0)\| < \delta,$$

then

$$\|y(t)\| < \varepsilon \qquad \text{for all} \qquad t \geqslant t_0 \geqslant 0$$

independently of the specific value of $t_0$. □

A theorem concerning absolutely integrable perturbations is frequently given in literature [**4**], Chapter II, in the following form.

THEOREM 4.4.5. *Let*
1) *system* (4.4.1) *be stable,*
2) $\int_0^\infty \|B(\tau)\|\, d\tau \leqslant \beta < \infty$,
3) $\underline{\lim}_{t\to\infty} \int_{t_0}^{t} \operatorname{Re} \operatorname{Sp} A(\tau)\, d\tau \geqslant \alpha > -\infty$.
*Then system* (4.4.10) *is also stable.*

PROOF. Let $X(t)$ be a fundamental matrix of (4.4.1); this matrix is nonsingular, and if we show that

$$|\operatorname{Det} X(t)| \geqslant c > 0, \qquad t \in \mathbb{R}_+,$$

then $X^{-1}$ will also be bounded on $\mathbb{R}_+$. Recall that the boundedness of $X(t)$ itself follows from the stability of the system. By the Liouville-Ostrogradskiĭ formula (1.0.5), we have

$$\operatorname{Det} X(t) = \operatorname{Det} X(t_0)e^{\int_{t_0}^{t} \operatorname{Sp} A(\tau)\, d\tau},$$

and by virtue of condition 3) of the theorem we have

$$|\operatorname{Det} X(t)| \geqslant |\operatorname{Det} X(t_0)|e^{\alpha} > 0.$$

We turn to the system (4.4.10) and show that any solution $y(t)$ of it is bounded on $\mathbb{R}_+$; this implies its stability. Let $t_0 \in \mathbb{R}_+$; we have

$$y(t) = X(t)X^{-1}(t_0)y(t_0) + \int_{t_0}^{t} X(t)X^{-1}(\tau)B(\tau)y(\tau)\,d\tau,$$

or

$$\|y(t)\| \leqslant c_1\|y(t_0)\| + c_2\int \|B(\tau)\|\|y(\tau)\|\,d\tau,$$

or

$$\|y(t)\| \leqslant c_1\|y(t_0)\|e^{c_2\int_{t_0}^{t}\|B(\tau)\|\,d\tau} \leqslant c_1e^{c_2\beta}\|y(t_0)\|. \quad \square$$

Compare the last two theorems. At first glance, the formulation of the second theorem seems more attractive, since it is easier to establish stability than uniform stability, and in order to check condition 3) of the theorem one has only to know the coefficients of the system. But, in fact, conditions 1) and 3) impose a stricter requirement on the system than that of uniform stability. Indeed, conditions 1) and 3) of the second theorem require the boundedness of the fundamental matrix as well as of the inverse to it, i.e., the stability not only of the system itself, but also of the adjoint one. Such stability is called *bounded stability*. It implies uniform stability since

$$\|X(t,s)\| = \|X(t)X^{-1}(s)\| \leqslant \|X(t)\|\|X^{-1}(s)\| \leqslant C,$$

but not conversely. For example, in the case of a constant matrix with negative eigenvalues, condition 3) of the theorem does not hold, but the statement (by Theorem 4.3.1) is valid and follows from the first two conditions. By virtue of the above, Theorem 4.4.5 guarantees for system (4.4.10) not just stability, but uniform stability; however, the requirements of the theorem are excessive.

We consider a condition for the perturbation to preserve uniform asymptotic stability (an analog of Theorem 4.3.3).

THEOREM 4.4.6. *Let*

1) *system* (4.4.1) *be uniformly asymptotically stable, i.e.,*

$$\|X(t,s)\| \leqslant Ke^{-\alpha(t-s)}, \qquad t \geqslant s \geqslant 0, \quad K > 0, \quad \alpha > 0,$$

2) $\|B(t)\| \leqslant \delta$ *for* $t \geqslant 0$.

*Then the Cauchy matrix* $Y(t,s)$ *of system* (4.4.10) *satisfies the inequality*

$$\|Y(t,s)\| \leqslant Ke^{-\beta(t-s)}, \qquad t \geqslant s \geqslant 0, \tag{4.4.11}$$

*where* $\beta = \alpha - \delta K$.

*If, moreover,* $\beta > 0$, *then system* (4.4.10) *is uniformly asymptotically stable.*

PROOF. Any solution $y(t)$ of system (4.4.10) satisfies the integral equation

$$y(t) = X(t,s)y(s) + \int_s^t X(t,s)B(\tau)y(\tau)\,d\tau, \qquad t \geqslant s \geqslant 0.$$

Therefore,

$$\|y(t)\| \leqslant Ke^{-\alpha(t-s)}\|y(s)\| + K\int_s^t e^{-\alpha(t-\tau)}\|B(\tau)\|\|y(\tau)\|\,d\tau.$$

Multiplying this inequality by $e^{\alpha t}$ and setting

$$u(t) = e^{\alpha t}\|y(t)\|,$$

we obtain

$$u(t) \leqslant u(s)K + K\int_s^t \|B(\tau)\|u(\tau)\,d\tau, \qquad t \geqslant s.$$

By the Gronwall-Bellman lemma (see Appendix), we have

$$u(t) \leqslant Ku(s)e^{K\delta(t-s)},$$

or

$$\|y(t)\| \leqslant K\|y(s)\|e^{(-\alpha+K\delta)(t-s)}.$$

Since $y(t) = Y(t,s)y(s)$, we have

$$\|Y(t,s)y(s)\| \leqslant K\|y(s)\|e^{-(\alpha-K\delta)(t-s)},$$

and, finally,

$$\left\|Y(t,s)\frac{y(s)}{\|y(s)\|}\right\| \leqslant Ke^{-(\alpha-K\delta)(t-s)}.$$

The last inequality implies the required estimate (4.4.11). □

## §5. On perturbations preserving the spectrum of a system. The principle of linear inclusion

As before, we consider the unperturbed system (4.4.1) and the perturbed system (4.4.10). Let $\sigma_\Gamma$ be the Grobman irregularity coefficient of system (4.4.1).

THEOREM 4.5.1 (Grobman [**18**]). *If*

(4.5.1) $$\chi[B(t)] < -\sigma_\Gamma,$$

*then the spectra of the perturbed and unperturbed systems coincide.*

PROOF. Recall the definition of $\sigma_\Gamma$. Let

$$X(t) = \{x_1(t), \dots, x_n(t)\}$$

be a fundamental matrix of system (4.4.1),

$$V(t) = [X^{-1}(t)]^* = \{v_1(t), \dots, v_n(t)\},$$

and

$$\chi[x_i] = \alpha_i, \qquad \chi[v_i] = \mu_i, \qquad i = 1, \dots, n, \qquad \sigma_\Gamma = \min_X \max_i\{\alpha_i + \mu_i\}.$$

It is shown in the monograph [**9**] that the irregularity coefficient is realized at normal fundamental matrices, and in what follows $X(t)$ is assumed to be normal.

Let us show that to any solution $y(t)$ of system (4.4.10) we can associate a solution $x(t)$ of system (4.4.1) with the same characteristic exponent. Let $\chi[y] = \beta$; we write $y(t)$ in the form

(4.5.2) $$y(t) = X(t)X^{-1}(0)y(0) + X(t)\int_0^t X^{-1}(\tau)B(\tau)y(\tau)\,d\tau.$$

Denote

$$z(t) = X^{-1}(t)B(t)y(t)$$

and represent

$$\int_0^t z(\tau)\,d\tau = a + u(t),$$

where $a = (a_1, \dots, a_n)^\top$ is a constant vector such that

$$\begin{aligned} a_i &= 0 && \Rightarrow \quad u_i = \int_0^t z_i(\tau)\,d\tau, && \text{if} \quad \chi[z_i] \geqslant 0, \\ a_i &= \int_0^\infty z_i(\tau)\,d\tau && \Rightarrow \quad u_i = \int_\infty^t z_i(\tau)\,d\tau, && \text{if} \quad \chi[z_i] < 0. \end{aligned}$$

This implies (Theorem 2.1.5) that

$$\chi[u] \leqslant \chi[z].$$

Note that

$$z_i(t) = (B(t)y(t), \bar{v}_i(t)),$$

where the bar means complex conjugation. Therefore,

$$\chi[z_i] \leqslant \chi[B] + \beta + \mu_i. \tag{4.5.3}$$

We write (4.5.2) in the form

$$y(t) = X(t)[X^{-1}(0)y(0) + a] + X(t)u(t),$$

or

$$y(t) = x(t) + w(t), \tag{4.5.4}$$

where $x(t)$ is a solution of system (4.4.1). Let us estimate $\chi[w(t)]$,

$$\begin{aligned} \chi[w_j(t)] &= \chi[X(t)u(t)]_j \\ &= \chi\left[\sum_{i=1}^n x_{ji}(t)u_i(t)\right] \\ &\overset{(4.5.3)}{\leqslant} \max_i\{\alpha_i + \mu_i\} + \beta + \chi[B] \\ &\overset{(4.5.1)}{<} \max_i\{\alpha_i + \mu_i\} - \sigma_\Gamma + \beta \\ &= \beta. \end{aligned}$$

Thus, $\chi[w] < \beta$, and from (4.5.4) we have $\chi[x] = \beta$.

Let us take a normal basis

$$Y(t) = \{y_1(t), \dots, y_n(t)\}$$

of system (4.4.10), associate to it the set

$$X(t) = \{x_1(t), \dots, x_n(t)\}$$

of solutions of system (4.4.1) with the same characteristic exponents, and show that $X(t)$ is also a normal basis.

Indeed, from (4.5.4) we have

$$x_i(t) = y_i(t) - w_i(t),$$

where

$$\chi[w_i] < \chi[y_i] = \chi[x_i].$$

It follows from the incompressibility of the basis $X(t)$ that it is normal, since in the converse case there would also exist a combination reducing the characteristic exponent for the basis $Y(t)$, which contradicts the assumption that it is normal. □

REMARK 4.5.1. The theorem certainly remains valid if we change $\sigma_\Gamma$ in (4.5.1) to

$$\sigma_\Lambda = \sum_{s=1}^{n} \alpha_s - \varliminf_{t\to\infty} \frac{1}{t}\int_{t_0}^{t} \operatorname{Re}\operatorname{Sp} A(\tau)\,d\tau,$$

since $\sigma_\Lambda \geqslant \sigma_\Gamma$. Here $\{\alpha_1, \dots, \alpha_n\}$ is the complete spectrum of system (4.4.1).

It is in this form that the theorem was independently proved by Bogdanov **[8]**. Neither of the two coefficients of irregularity can be found explicitly; however, the estimate for $\sigma_\Lambda$ is easier to obtain by using any estimate of the characteristic exponents (see the next section of this chapter).

Many results obtained for systems under a linear perturbation can be extended to the case of an almost linear perturbation by means of the following theorem.

Consider a system

$$\dot{y} = A(t)y + f(t, y), \tag{4.5.5}$$

where

$$A \in C(\mathbb{R}_+), \qquad \sup_{t\geqslant 0} \|A(t)\| < \infty,$$

$$\|f(t, y)\| \leqslant g(t)\|y\| \tag{4.5.6}$$

for a continuous function $g(t)$, bounded for $t \geqslant 0$.

THEOREM 4.5.2 (the principle of linear inclusion **[9]**). *For any solution $y(t)$ of system (4.5.5) there exists a linear system*

$$\dot{z} = A(t)z + B_y(t)z, \tag{4.5.7}$$

$$\|B_y(t)\| \leqslant g(t), \tag{4.5.8}$$

*whose solution is $z = y(t)$.*

PROOF. Let $y(t)$ be a nontrivial solution of system (4.5.5). By virtue of the condition (4.5.6), $y_0(t) \equiv 0$ is a solution of (4.5.5). By the uniqueness theorem,

$$\|y(t)\| \neq 0, \qquad t \in \mathbb{R}_+.$$

Let us set

$$G(t, s) = \frac{1}{\|y(t)\|^2}(z, y(t)) f(t, y(t));$$

$G(t, z)$ is a vector whose components are linear combinations of the components of the vector $z$; this enables us to write

$$G(t, z) = B_y(t)z, \qquad B_y(t) \text{ is a matrix.}$$

Let us show that $y(t)$ is a solution of system (4.5.7) for this choice of the matrix $B_y(t)$,

$$\dot{y}(t) \equiv A(t)y(t) + f(t, y(t)) = A(t)y(t) + G(t, y(t)) = A(t)y(t) + B_y(t)y(t).$$

It remains to verify (4.5.8):

$$\|B_y(t)z\| = \|G(t,z)\| \leqslant \frac{\|z\|\|y\|}{\|y\|^2}\|f(t,y)\| \leqslant g(t)\|z\|,$$

or

$$\left\|B_y(t)\frac{z}{\|z\|}\right\| \leqslant g(t). \quad \square$$

REMARK 4.5.1′. Different systems (4.5.7) correspond to different solutions of system (4.5.5); therefore, the principle is applicable to each solution separately.

In spite of the above remark, this approach often allows one to investigate general properties of solutions of (4.5.5). For example, assuming that $g(t)$ is small in a certain sense and using the theorems from this chapter, we can make conclusions about the character of the solutions of systems (4.5.7) and, consequently, of those of the initial system. For example, we have the following statement.

THEOREM 4.5.3. *The set of characteristic exponents of system* (4.4.5), *where* $\chi[g] < -\sigma_\Gamma$, *belongs to the spectrum of the unperturbed system.*

PROOF. Let $y(t)$ be a solution of system (4.5.5) and $\chi[y] = \beta$; at the same time $y(t)$ is a solution of some system (4.5.7) whose spectrum, by Theorem 4.5.1, coincides with the spectrum of the unperturbed system; therefore, $\beta$ belongs to this spectrum.□

We shall give another example of application of the principle of linear inclusion.

THEOREM 4.5.4. *If all the solutions of system* (4.5.5) *under any perturbation* $f(t,y)$ *with* $g(t)$ *of a fixed smallness admit the estimate*

$$\|y(t)\| \leqslant \|y(0)\|H(t), \tag{4.5.9}$$

*then all the solutions of the system*

$$\dot{v} = -A^*(t)v + \varphi(t,v), \qquad \|\varphi(t,v)\| \leqslant \tilde{g}(t)\|v\|, \tag{4.5.10}$$

*with* $\tilde{g}(t)$ *of the same smallness admit the estimate*

$$\|v(t)\| \geqslant \|v(0)\|\frac{1}{H(t)}. \tag{4.5.11}$$

PROOF. Let us take a solution $v(t)$ of system (4.5.10). It satisfies some linear system

$$\dot{z} = -A^*(t)z + B_v(t)z, \qquad \|B_v(t)\| \leqslant \tilde{g}(t),$$

and the system

$$\dot{w} = A(t)w - B_v^*(t)w \tag{4.5.12}$$

adjoint to it is such that

$$\|B_v^*(t)\| = \|B_v(t)\| \leqslant \tilde{g}(t).$$

By the condition of the theorem, any solution $w(t)$ of system (4.5.12) satisfies the estimate (4.5.9). Now let us choose $w(t)$ such that

$$w(0) = v(0).$$

The functions $v(t)$ and $w(t)$ are solutions of adjoint systems. Therefore,

$$(w(t), v(t)) \equiv \text{const}, \qquad t \geqslant 0.$$

Hence,

$$(w(t), v(t)) = (w(0), v(0)) = (v(0), v(0)) = \|v(0)\|^2,$$

or

$$\|v(0)\|^2 \leqslant \|w(t)\|\|v(t)\| \overset{(4.5.9)}{\leqslant} \|w(0)\|H(t)\|v(t)\|.$$

From the last inequality we obtain (4.5.11). □

## §6. Growth estimates of the solutions of linear systems in terms of the coefficients

Consider a system

$$\dot{x} = A(t)x, \tag{4.6.1}$$

where

$$x \in \mathbb{C}^n, \qquad A \in C(\mathbb{R}_+), \qquad \sup_{t \geqslant 0} \|A(t)\| < \infty.$$

In this section we deal with six different coefficient criteria for the growth of the solutions as $t \to \infty$. The first estimate of this sort is due to Lyapunov. By means of this estimate it has been proved that the characteristic exponents of linear systems are finite in the case when the coefficients are bounded.

**I. Lyapunov's estimate.** For any solution $x(t)$ of system (4.6.1), the inequality

$$\|x(t_0)\|e^{-\int_{t_0}^t \|A(\tau)\|\,d\tau} \leqslant \|x(t)\| \leqslant \|x(t_0)\|e^{\int_{t_0}^t \|A(\tau)\|\,d\tau}, \qquad t \geqslant t_0 \geqslant 0, \tag{4.6.2}$$

is valid; the right-hand side of this inequality was already repeatedly used. Recall the derivation. We pose the Cauchy problem

$$x(t_0) \in \mathbb{C}^n, \qquad t_0 \in \mathbb{R}_+,$$

for system (4.6.1). This problem is equivalent to the integral equation

$$x(t) = x(t_0) + \int_{t_0}^t A(\tau)x(\tau)\,d\tau, \qquad t, t_0 \in \mathbb{R}_+,$$

whence

$$\|x(t)\| \leqslant \|x(t_0)\| + \left| \int_{t_0}^t \|A(\tau)\|\|x(\tau)\|\,d\tau \right|,$$

and by the Gronwall-Bellman lemma we obtain (4.6.2).

**II. Bogdanov's estimate.**

THEOREM 4.6.1 (Bogdanov **[8]**). *For any solution $x(t)$ of system* (4.6.1) *the estimate*

$$\|x(0)\|e^{-\frac{1}{2}\int_0^t \sum_{i,j=1}^n |a_{ij}+a_{ji}|\,d\tau} \leqslant x(t) \leqslant \|x(0)\|e^{\frac{1}{2}\int_0^t \sum_{i,j=1}^n |a_{ij}+a_{ji}|\,d\tau}, \tag{4.6.3}$$

*where*

$$\|x(t)\| \quad \textit{is the Euclidean norm}, \qquad \operatorname{Im} A(t) \equiv 0, \qquad t \in \mathbb{R}_+,$$

*is valid.*

PROOF. Take a nontrivial solution

$$x(t) = (x_1(t), \dots, x_n(t))^\top$$

of system (4.6.1); for it we have

$$\frac{dx_i}{dt} = \sum_{j=1}^{n} a_{ij}(t)x_j, \qquad i = 1, \dots, n,$$

or

$$x_i\dot{x}_i = \sum_{j=1}^{n} a_{ij}(t)x_i x_j.$$

Summing up the last identities, we obtain

$$\frac{d}{dt}\sum_{i=1}^{n} x_i^2 = 2\sum_{i,j=1}^{n} a_{ij}(t)x_i x_j = \sum_{i,j=1}^{n}[a_{ij}(t) + a_{ji}(t)]x_i x_j.$$

Further, we have

$$\begin{aligned} \left|\frac{d}{dt}\|x\|^2\right| &\leqslant \sum_{i,j=1}^{n} |a_{ij} + a_{ji}| \frac{x_i^2 + x_j^2}{2} \\ &= \sum_{i=1}^{n}\left(\sum_{j=1}^{n} |a_{ij} + a_{ji}| x_i^2\right) \\ &\leqslant \left(\sum_{i=1}^{n} x_i^2\right)\sum_{l,s=1}^{n} |a_{ls} + a_{sl}|. \end{aligned}$$

From the last inequality we have

$$-\sum_{l,s=1}^{n} |a_{ls} + a_{sl}| \|x\|^2 \leqslant \frac{d\|x\|^2}{dt} \leqslant \sum_{l,s=1}^{n} |a_{ls} + a_{sl}| \|x\|^2.$$

Dividing this result by $\|x\|^2$ and integrating, we obtain

$$-\int_0^t \sum_{l,s=1} |a_{ls} + a_{sl}|\, d\tau \leqslant 2(\ln\|x(t)\| - \ln\|x(0)\|) \leqslant \int_0^t \sum_{l,s=1}^{n} |a_{ls} + a_{sl}|\, d\tau;$$

this straightforwardly implies (4.6.3). □

Note that both these results have a serious shortcoming, i.e., the upper bound for the characteristic exponents is positive. Indeed, from the inequalities (4.6.2) we have

(4.6.4) $$-\underline{\lim}_{t\to\infty} \frac{1}{t}\int_{t_0}^{t} \|A(\tau)\|\, d\tau \leqslant \chi[x] \leqslant \overline{\lim}_{t\to\infty} \frac{1}{t}\int_{t_0}^{t} \|A(\tau)\|\, d\tau.$$

From Bogdanov's estimate (4.6.3) we have

(4.6.5) $$-\underline{\lim}_{t\to\infty} \frac{1}{2t}\int_0^t \sum_{i,j=1}^{n} |a_{ij} + a_{ji}|\, dt \leqslant \chi[x] \leqslant \overline{\lim}_{t\to\infty} \frac{1}{2t}\int_0^t \sum_{i,j=1}^{n} |a_{ij} + a_{ji}|\, d\tau.$$

Thus, inequalities (4.6.4) and (4.6.5) show that, in the case of an arbitrary matrix $A$, the estimates, generally speaking, are crude. Exceptions are the cases when $A \equiv 0$ and when $A(t)$ is skew-symmetric; both estimates in the first case and (4.6.3) in the second reflect the real state of affairs:

$$\|x(t)\| = \text{const}, \qquad t \in \mathbb{R}_+.$$

Now let us consider more precise estimates.

**III. Vazhevskiĭ's estimate.** This result does not require the assumption that $\|A(t)\|$ be bounded.

THEOREM 4.6.2 (Vazhevskiĭ **[19]**). *For any solution $x(t)$ of system* (4.6.1) *the inequality*

$$\|x(0)\|e^{\int_0^t \lambda(u)\,du} \leqslant \|x(t)\| \leqslant \|x(0)\|e^{\int_0^t \Lambda(u)\,du} \tag{4.6.6}$$

*is valid, where $\|x(t)\|$ is the Euclidean norm, and $\lambda(t)$ and $\Lambda(t)$ are the smallest and the greatest eigenvalues of the matrix*

$$A^H(t) = [A(t) + A^*(t)]/2.$$

PROOF. Let us take a nontrivial solution $x(t)$ of system (4.6.1); for it $\|x\|^2 = x^*x$. Therefore, we have

$$\frac{d\|x\|^2}{dt} = x^*\frac{dx}{dt} + \frac{dx^*}{dt}x = x^*A(t)x + x^*A^*(t)x = 2x^*A^H(t)x.$$

The matrix $A^H$ is Hermitian; hence, it is unitarily similar to the diagonal matrix **[16]**

$$D = \operatorname{diag}[\lambda_1(t), \dots, \lambda_n(t)].$$

Let $U(t)$ be such that

$$U^*(t)U(t) = E$$

and

$$A^H(t) = U^*(t)D(t)U(t);$$

hence

$$x^*A^Hx = x^*U^*DUx = y^*Dy = \sum_{j=1}^n \lambda_j y_j \bar{y}_j, \tag{4.6.7}$$

where $y = U(t)x$; therefore, $\|y\| = \|x\|$. Let

$$\lambda(t) = \min_i\{\lambda_i(t)\}, \qquad \Lambda(t) = \max_i\{\lambda_i(t)\}.$$

From (4.6.7) we have

$$\lambda(t)\|x\|^2 \leqslant x^*A^H(t)x \leqslant \Lambda(t)\|x\|^2,$$

or

$$2\lambda(t)\|x\|^2 \leqslant \frac{d\|x\|^2}{dt} \leqslant 2\Lambda(t)\|x\|^2.$$

Integrating the last inequality, we obtain (4.6.6). □

REMARK 4.6.1. From Vazhevskiĭ's theorem we have

$$\varlimsup_{t\to\infty}\frac{1}{t}\int_0^t \lambda(u)\,du \leqslant \chi[x] \leqslant \varlimsup_{t\to\infty}\frac{1}{t}\int_0^t \Lambda(\tau)\,d\tau. \tag{4.6.8}$$

The right-hand side of the inequality (4.6.8) may be negative.

EXAMPLE 4.6.1. Consider system (4.6.1) with the matrix

$$A=\begin{pmatrix}-1 & t\\ -t & -4\end{pmatrix},\qquad \text{where}\qquad A^H=\frac{1}{2}(A+A^*)=\begin{pmatrix}-1 & 0\\ 0 & -4\end{pmatrix}.$$

Therefore, $\Lambda=-1$, $\lambda=-4$, and the solutions of the system

$$\dot{x}_1=-x_1+tx_2,$$
$$\dot{x}_2=-tx_1-4x_2$$

have the estimate

$$e^{-4t}\|x(0)\|\leqslant\|x(t)\|\leqslant\|x(0)\|e^{-t};$$

we note that the coefficients of the system are unbounded and the arguments from §3 and §4, as well as the previous estimates, are inapplicable to it.

**IV. Lozinskiĭ's estimates.** In this subsection we deal with Lozinskiĭ's logarithmic norm and its application to estimates for solutions of linear systems.

DEFINITION 4.6.1 **[24]**. The number $\gamma(A)$ defined as

$$\gamma(A)=\lim_{h\to0+}\frac{\|E+hA\|-1}{h} \tag{4.6.9}$$

is called the *logarithmic norm* of the matrix $A$.

Note that $\gamma(A)$ depends on the choice of the matrix norm and is defined for any norm such that $\|E\|=1$. The term is conditional since this norm does not have the properties of an ordinary norm; for example, it may be negative.

EXAMPLE 4.6.2. For the matrix

$$A=\begin{pmatrix}-5 & 1\\ 2 & -3\end{pmatrix}\qquad\text{we have}\qquad \gamma_I(A)=-1.$$

Here and below the subscripts *I*, *II*, *III* indicate which of the norms we use is chosen (see Chapter I, §2).

Note the properties of the logarithmic norm.

1. $\gamma(A+B)\leqslant\gamma(A)+\gamma(B)$. Let us prove this. We have

$$\|E+h(A+B)\|\leqslant\left\|\frac{1}{2}E+hA\right\|+\left\|\frac{1}{2}E+hB\right\|=\frac{1}{2}[\|E+2hA\|+\|E+2hB\|].$$

Subtract 1 from both sides and divide by $h$; then pass to the limit.

2. $\gamma(\alpha A)=\alpha\gamma(A)$, $\alpha\in\mathbb{R}_+$ (this is verified by means of (4.6.9)).

3. $\gamma(A)\leqslant\|A\|$.

Indeed,

$$\|E+hA\|-1=\|E+hA\|-\|E\|\leqslant\|E\|+\|hA\|-\|E\|=h\|A\|;$$

it remains to divide by $h$ and to pass to the limit.

4. $\gamma(A) - \gamma(B) \leqslant \|A - B\|$; this follows from

$$\gamma(A) \leqslant \gamma(B) + \gamma(A - B) \leqslant \gamma(B) + \|A - B\|.$$

We give the values of the logarithmic norms for the three matrix norms indicated:

$$\gamma_I(A) = \max_\mu \left\{ \operatorname{Re} a_{\mu\mu} + \sum_{\eta \neq \mu} |a_{\mu\eta}| \right\},$$

$$\gamma_{II}(A) = \max_\eta \left\{ \operatorname{Re} a_{\eta\eta} + \sum_{\mu \neq \eta} |a_{\mu\eta}| \right\},$$

$$\gamma_{III}(A) = \text{ the greatest eigenvalue of } (A + A^*)/2.$$

The calculations of these norms are carried out straightforwardly. For example,

$$\begin{aligned} \|E + hA\|_I &= \max_\mu \sum_{\eta=1}^{n} |(E + hA)_{\mu\eta}| \\ &= \max_\mu \left\{ |1 + h a_{\mu\mu}| + \sum_{\eta \neq \mu} h |a_{\mu\eta}| \right\} \\ &= \max_\mu \left\{ 1 + h \operatorname{Re} a_{\mu\mu} + O(h^2) + \sum_{\eta \neq \mu} h |a_{\mu\eta}| \right\}. \end{aligned}$$

Subtracting 1 and dividing by $h$, we pass to the limit and obtain $\gamma_I(A)$.

Now we turn to system (4.6.1). Let $X(t, s)$ be its Cauchy matrix.

THEOREM 4.6.3 (Lozinskiĭ). *For any matrix norm the following estimate is valid:*

$$\|X(t, s)\| \leqslant e^{\int_s^t \gamma(A(\tau))\, d\tau}, \qquad t \geqslant s \geqslant 0. \tag{4.6.10}$$

PROOF. Let $t_k = s + (k/N)(t - s)$; however, the division of the interval $[s, t]$ does not necessarily have to be uniform, only the maximal length of the subintervals must not exceed $h$. We have

$$X(t, s) = X(t_N) X^{-1}(t_{N-1}) X^{-1}(t_{N-2}) \cdots X(t_1) X^{-1}(s),$$

or

$$X(t, s) = \prod_{k=N}^{1} X(t_k, t_{k-1}).$$

Recall that

$$X(t_k, t_{k-1}) \equiv \underset{t_{k-1}}{\overset{t_k}{\Omega}} A$$

and turn to the expansion (1.2.7):

$$\begin{aligned} X(t_k, t_{k-1}) &= E + \int_{t_{k-1}}^{t_k} A(u)\, du + \int_{t_{k-1}}^{t_k} A(u_1)\, du_1 \int_{t_{k-1}}^{u_1} A(u_2)\, du_2 + \cdots \\ &= E + \int_{t_{k-1}}^{t_k} A(u)\, du + O(h^2) = E + hA(t_{k-1}) + O(h^2). \end{aligned}$$

We pass to the estimate

$$\begin{aligned}
\|X(t,s)\| &\leqslant \prod_{k=1}^{N} \|E + hA(t_{k-1}) + O(h^2)\| \\
&\leqslant \prod_{k=1}^{N} (\|E + hA(t_{k-1})\| + O(h^2)) \overset{(4.6.9)}{=} \prod_{k=1}^{N} [1 + h\gamma(A(t_{k-1})) + o(h)] \\
&= \prod_{k=1}^{N} e^{\ln[1+h\gamma(A(t_{k-1}))+o(h)]} = e^{\sum_{k=1}^{N}[h\gamma(A(t_{k-1}))+o(h)]} \\
&= e^{\sum_{k=1}^{N} h\gamma(A(t_{k-1}))+No(h)} = e^{\sum_{k=1}^{N} h\gamma(A(t_{k-1}))+((t-s)/h)o(h)}.
\end{aligned}$$

Passing to the limit as $h \to 0+$, we obtain the estimate (4.6.10). □

It is possible to obtain analogous estimates for the Cauchy matrix from below [**9**]:

$$\|X(t,s)\| \geqslant \exp\left[\int_s^t \lim_{h\to 0-} \frac{\|E + hA(\tau)\| - 1}{h}\, d\tau\right]. \tag{4.6.11}$$

Combining the estimates (4.6.10) and (4.6.11), we write them for the norms $I$, $II$, $III$, respectively:

1. $$\exp\left(\int_s^t \min_\mu \left(\operatorname{Re} a_{\mu\mu} - \sum_{\eta\neq\mu} |a_{\mu\eta}|\right) d\tau\right) \leqslant \|X(t,s)\| \leqslant \exp\left(\int_s^t \max_\mu \left(\operatorname{Re} a_{\mu\mu} + \sum_{\eta\neq\mu} |a_{\mu\eta}|\right) d\tau\right), \tag{4.6.12}$$

2. $$\exp\left(\int_s^t \min_\eta \left(\operatorname{Re} a_{\eta\eta} - \sum_{\mu\neq\eta} |a_{\mu\eta}|\right) d\tau\right) \leqslant \|X(t,s)\| \leqslant \exp\left(\int_s^t \max_\eta \left(\operatorname{Re} a_{\eta\eta} + \sum_{\mu\neq\eta} |a_{\mu\eta}|\right) d\tau\right), \tag{4.6.13}$$

3. for the third norm we obtain Vazhevskiĭ's estimate (4.6.6).

REMARK 4.6.2. The estimates (4.6.12) and (4.6.13) given here are not invariant under Lyapunov transformations. We can try to improve them by passing from the matrix $A(t)$ to the matrix

$$B(t) = L^{-1}(t)A(t)L(t) - L^{-1}\dot{L}(t),$$

where $L(t)$ is a Lyapunov matrix. This process is quite cumbersome. Let us illustrate Lozinskiĭ's result by examples.

Note that

$$\begin{aligned}\delta(1+t-\tau)^{n-1}(t-\tau) &\leqslant \delta^{1/(n+1)}\delta^{n/(n+1)}(1+t-\tau)^n \\ &= \delta^{1/(n+1)}e^{n\ln[\delta^{1/n+1}(1+t-\tau)]} \\ &\leqslant \delta^{1/(n+1)}e^{n\delta^{1/(n+1)}(1+t-\tau)}.\end{aligned} \tag{4.6.18}$$

Thus, by virtue of the conditions (4.6.16) and (4.6.17), we write

$$\|x(t)\| \leqslant D(1+t)^{n-1}e^{\gamma t}\|x(0)\| + \int_0^t D(1+t-\tau)^{n-1}e^{\gamma(t-\tau)}\delta(t-\tau)\|x(\tau)\|\,d\tau,$$

and, taking into account (4.6.18), we have

$$\|x(t)\| \leqslant D(1+t)^{n-1}e^{\gamma t}\|x(0)\| + \int_0^t De^{\gamma(t-\tau)}\delta^{1/(n+1)}e^{n\delta^{1/(n+1)}(1+t-\tau)}\|x(\tau)\|\,d\tau.$$

Divide the last equality by

$$\varphi(t) = (1+t)^{n-1}e^{(\gamma+n\delta^{1/(n+1)})t} > 0. \tag{4.6.19}$$

Then

$$\frac{\|x(t)\|}{\varphi(t)} \leqslant De^{-n\delta^{1/(n+1)}t}\|x(0)\| + \int_0^t \frac{D\delta^{1/(n+1)}}{(1+t)^{n-1}}e^{n\delta^{1/(n+1)}}e^{-(\gamma+n\delta^{1/(n+1)})\tau}\|x(\tau)\|\,d\tau;$$

therefore,

$$\frac{\|x(t)\|}{\varphi(t)} \leqslant D\|x(0)\| + \int_0^t De^{n\delta^{1/(n+1)}}\delta^{1/(n+1)}\frac{\|x(\tau)\|}{\varphi(\tau)}\,d\tau.$$

By the Gronwall-Bellman formula, we have

$$\frac{\|x(t)\|}{\varphi(t)} \leqslant D\|x(0)\|e^{\int_0^t D_1\delta^{1/(n+1)}\,d\tau}, \qquad D_1 = De^{n\delta^{1/(n+1)}},$$

or, taking into account (4.6.19), we obtain

$$\|x(t)\| \leqslant D\|x(0)\|(1+t)^{n-1}e^{(\gamma+n\delta^{1/(n+1)})t}e^{D_1\delta^{1/(n+1)}t}. \tag{4.6.20}$$

Thus, we have proven the following statement.

THEOREM 4.6.4. *Let system* (4.6.1) *be such that*

$$\|A(t_1) - A(t_2)\| \leqslant \delta|t_1 - t_2|,$$

*where $\delta$ is sufficiently small. Then*

1) *for any solution $x(t)$ the estimate* (4.6.20) *is valid,*
2) *we have*

$$\chi[x] \leqslant \gamma + d\delta^{1/(n+1)}, \tag{4.6.21}$$

*where $\gamma$ is defined by formula* (4.6.14) *and $d = n + De^{n\delta^{1/(n+1)}}$. Here $n$ is the order of the system and $D$ is the constant from the estimate* (1.3.11).

REMARK 4.6.3. The attainability of the estimate (4.6.21) was proved for systems of any order in **[23]**. This means that there exist systems whose greatest characteristic exponent is not less than

$$\gamma + c_0\delta^{1/(n+1)},$$

where $c_0$ is a constant independent of $\delta$.

**VI. Yakubovich's estimate for the characteristic exponents of systems with periodic coefficients [39].** Let the matrix $A(t)$ in system (4.6.1) be such that

$$A(t) = A(t+\omega), \qquad t \in \mathbb{R}.$$

We introduce the set of matrices $G(t)$ satisfying the following conditions for $t \in \mathbb{R}$:

1. $G^*(t) = G(t)$,
2. $G(t+\omega) = G(t)$,
3. $G \in C^1(\mathbb{R})$,
4. $(G(t)a, a) > 0$, where $a \in \mathbb{C}^n$ is arbitrary, $\|a\| \neq 0$, $t \in \mathbb{R}$.

For example, for any $\omega$-periodic and continuously differentiable matrix $F(t)$, the matrix

$$G(t) = F^*(t)F(t)$$

satisfies the conditions given above.

Let $\rho$ be a multiplier of system (4.6.1). By Theorem 1.4.1, the solution $x = e^{\lambda t}\varphi(t)$, where

$$\varphi(t+\omega) = \varphi(t) \qquad \text{and} \qquad \lambda = \frac{1}{\omega}\operatorname{Ln}\rho,$$

corresponds to it. Consider the form

$$\xi(t) = (G(t)x, x). \tag{4.6.22}$$

We take our normal solution as $x(t)$; then

$$\xi(t) = e^{(\lambda+\bar{\lambda})t}(G\varphi, \varphi) = e^{2t\operatorname{Re}\lambda}(G\varphi, \varphi),$$

whence

$$\ln\xi(t) = 2t\operatorname{Re}\lambda + \ln(G\varphi, \varphi). \tag{4.6.23}$$

Note that

$$\ln(G(t)\varphi(t), \varphi(t))$$

is a periodic and, consequently, a bounded function; therefore, dividing (4.6.23) by $2t$, in the limit we have

$$\operatorname{Re}\lambda = \frac{1}{2}\lim_{t\to\infty}\frac{1}{t}\ln\xi(t). \tag{4.6.24}$$

Starting with this, we obtain estimates for $\operatorname{Re}\lambda$. Differentiate the form (4.6.22) along the trajectories of system (4.6.1):

$$\dot{\xi} = (\dot{G}x, x) + (GAx, x) + (Gx, Ax) = (Qx, x),$$

where

$$Q = \dot{G} + GA + A^*G \qquad (Q^* = Q). \tag{4.6.25}$$

Consider the equation

$$\operatorname{Det}(Q - qG) = 0. \tag{4.6.26}$$

Let $q_1(t)$ and $q_2(t)$ be its smallest and greatest roots; note that they are real and $\omega$-periodic. Indeed, introduce the matrix $G^{1/2}$. The matrix $G$ is Hermitian; therefore, there exists a unitary matrix $U$ such that $G = U^*DU$, where $D$ is real diagonal [**16, 27**]. We write

$$G = U^*D^{1/2}UU^*D^{1/2}U \equiv G^{1/2}G^{1/2}.$$

Note that $G^{1/2}$ is Hermitian; moreover,

$$\mathrm{Det}[(G^{-1/2})^*[Q - qG]G^{-1/2}] = \mathrm{Det}[(G^{-1/2})^*QG^{1/2} - qE],$$

whence we obtain the equivalence of (4.6.26) and

$$\mathrm{Det}[(G^{-1/2})^*QG^{-1/2} - qE] = 0. \tag{4.6.27}$$

The matrix

$$H = (G^{-1/2})^*QG^{-1/2}$$

is Hermitian; hence [**16, 27**], the inequality

$$h_{\min}(y, y) \leqslant (Hy, y) \leqslant h_{\max}(y, y)$$

holds for any vector $y = \mathbb{C}^n$, where

$$h_{\min} = \min_i\{h_1(t), \dots, h_n(t)\}, \qquad h_{\max} = \max_i\{h_1(t), \dots, h_n(t)\},$$

and $h_1(t), \dots, h_n(t)$ are the eigenvalues of $H(t)$. By virtue of this, (4.6.27) implies

$$q_1(t)(y, y) \leqslant (QG^{-1/2}y, G^{-1/2}y) \leqslant q_2(t)(y, y),$$

and, setting $y = G^{1/2}x$, we obtain

$$q_1(t)(G^{1/2}x, G^{1/2}x) \leqslant (Qx, x) \leqslant q_2(t)(G^{1/2}x, G^{1/2}x),$$

or

$$q_1(t)(Gx, x) \leqslant (Qx, x) \leqslant q_2(t)(Gx, x),$$

or

$$q_1(t)\xi(t) \leqslant \dot{\xi}(t) \leqslant q_2(t)\xi(t). \tag{4.6.28}$$

The inequalities (4.6.28) give a two-sided estimate of the function $\xi(t)$:

$$\int_0^t q_1(\tau)\,d\tau \leqslant \ln\xi(t) - \ln\xi(0) \leqslant \int_0^t q_2(\tau)\,d\tau. \tag{4.6.29}$$

For any $\omega$-periodic function $q(t)$ we have

$$\int_0^t q(\tau)\,d\tau = \frac{t}{\omega}\int_0^t q(\tau)\,d\tau + r(t),$$

where $r(t)$ is $\omega$-periodic. Thus,

$$\lim_{t\to\infty}\frac{1}{t}\int_0^t q(\tau)\,d\tau = \frac{1}{\omega}\int_0^{\omega} q(\tau)\,d\tau.$$

Dividing inequality (4.6.29) by $2t$, we pass to the limit, and, by virtue of (4.6.24), we obtain

$$\frac{1}{2\omega}\int_0^\omega q_1(\tau)\,d\tau \leqslant \operatorname{Re}\lambda \leqslant \frac{1}{2\omega}\int_0^\omega q_2(\tau)\,d\tau. \tag{4.6.30}$$

Our reasoning shows that the result essentially depends on the choice of $G(t)$. By an appropriate choice of $G(t)$, the estimate (4.6.30) can be made as precise as desired.

THEOREM 4.6.5 (Yakubovich [**39**]). *Let*

$$\alpha_1 \leqslant \alpha_2 \leqslant \cdots \leqslant \alpha_n$$

*be the spectrum of system* (4.6.1). *For any $\varepsilon > 0$ there exists a matrix $G(t)$ satisfying the conditions formulated above such that*

$$\begin{gathered}\frac{1}{2\omega}\int_0^\omega q_1(\tau)\,d\tau \leqslant \alpha_1 < \frac{1}{2\omega}\int_0^\omega q_1(\tau)\,d\tau + \varepsilon,\\ \frac{1}{2\omega}\int_0^\omega q_2(\tau)\,d\tau - \varepsilon < \alpha_n \leqslant \frac{1}{2\omega}\int_0^\omega q_2(\tau)\,d\tau.\end{gathered} \tag{4.6.31}$$

*If the matrix $A(t)$ is real, then the matrix $G(t)$ can also be chosen real.*

We omit the proof (the reader can find it in the monograph [**39**]).

Let us consider the case of a second order real system in more detail. Here the matrix $G(t)$ has the form

$$G(t) = \begin{pmatrix} a(t) & b(t) \\ b(t) & c(t)\end{pmatrix},$$

and the conditions for it are reduced to the requirements that the elements of the matrix be continuously differentiable, $\omega$-periodic, and

$$a(t) > 0, \quad c(t) > 0, \quad a(t)c(t) - b^2(t) > 0, \quad 0 \leqslant t \leqslant \omega.$$

The matrix $Q(t)$ is defined by (4.6.25), and equation (4.6.26) can be written as

$$\operatorname{Det}(QG^{-1} - qE) = 0,$$

or

$$q^2 - \operatorname{Sp}(QG^{-1})q + \operatorname{Det}(QG^{-1}) = 0, \tag{4.6.32}$$

$$\operatorname{Sp} QG^{-1} = \operatorname{Sp}(\dot G G^{-1} + GAG^{-1} + A^*) = \operatorname{Sp}\dot G G^{-1} + 2\operatorname{Sp} A.$$

The matrix $G(t)$ satisfies some matrix equation $\dot G = B(t)G$; thus,

$$B(t) = \dot G G^{-1}.$$

Therefore, by the Ostrogradskiĭ-Liouville formula, we have

$$\operatorname{Det} G(t) = \operatorname{Det} G(0) e^{\int_0^t \operatorname{Sp} B(\tau)\,d\tau},$$

or

$$\operatorname{Sp}\dot G G^{-1} = \frac{d}{dt}\ln \operatorname{Det} G(t).$$

The solution of equation (4.6.32) has the form

$$q_{1,2} = \frac{1}{2}(\operatorname{Sp}\dot G G^{-1} + 2\operatorname{Sp} A) \pm \left(\frac{1}{4}[\operatorname{Sp}\dot G G^{-1} + 2\operatorname{Sp} A]^2 - \operatorname{Det} QG^{-1}\right)^{1/2}.$$

Hence (since $\int_0^\omega \operatorname{Sp}\dot{G}G^{-1}\,d\tau = 0$), by Theorem 4.6.5, we obtain
(4.6.33)

$$\alpha_{1,2} = \frac{1}{2\omega}\left[\int_0^\omega \operatorname{Sp} A(\tau)\,d\tau \pm \inf_G \int_0^\omega \left(\frac{1}{4}[\operatorname{Sp}\dot{G}G^{-1} + 2\operatorname{Sp} A]^2 - \operatorname{Det} QG^{-1}\right)^{1/2} d\tau\right];$$

this implies the estimates (4.6.31).

Now we apply these arguments to Malkin's example considered above.

EXAMPLE 4.6.6 (Malkin). Consider the system

$$\dot{x}_1 = \mu(-1 + 2\sin t)x_1 + \mu x_2,$$
$$\dot{x}_2 = \mu x_1 - x_2.$$

For $G$ we take $E$; therefore,

$$Q = A + A^* = 2\begin{pmatrix} \mu(-1 + 2\sin t) & \mu \\ \mu & -1 \end{pmatrix},$$
$$\operatorname{Det} Q = -4[\mu(-1 + 2\sin t) + \mu^2].$$

Our goal is to find out for which $\mu$ Malkin's system is asymptotically stable, i.e., when $\alpha_2 < 0$. From formulas (4.6.33) we have

$$\alpha_2 \leqslant \frac{1}{2}\left[-1 - \mu + \frac{1}{2\pi}\int_0^{2\pi} [(1 + \mu(-1 + 2\sin t))^2 + 4\mu^2]^{1/2}\,dt\right].$$

Let us compare this estimate with Example 4.6.4. By means of Yakubovich's method we have reduced the problem to the same inequality as by means of Lozinskiĭ's method. In that case it was solved numerically, and here the integral can be simplified by means of the Cauchy-Bunyakovskiĭ inequality:

$$\frac{1}{\omega}\int_0^\omega \sqrt{\psi(t)}\,dt \leqslant \left(\frac{1}{\omega}\int_0^\omega \psi(t)\,dt\right)^{1/2}.$$

Therefore, we have

$$\alpha_2 \leqslant \frac{1}{2}\left(-1 - \mu + \sqrt{1 - 2\mu + 7\mu^2}\right).$$

Thus,

$$0 < \mu < 2/3.$$

CHAPTER V

# On the Variation of Characteristic Exponents Under Small Perturbations of Coefficients

In Chapter V we study the problem of the influence of small perturbations of the coefficients on the stability of linear systems, as well as on the change of the spectrum of a system under such perturbations. These problems are closely connected, but the latter requires more subtle methods, since it raises the question of change of all the elements of the spectrum.

Consider a system

$$\dot{x} = A(t)x, \tag{5.0.1}$$

$$A \in C(\mathbb{R}_+), \qquad \sup_{t\in\mathbb{R}_+} \|A(t)\| \leqslant M, \qquad x \in \mathbb{C}^n, \tag{5.0.2}$$

with spectrum

$$-\infty < \lambda_1 \leqslant \lambda_2 \leqslant \cdots \leqslant \lambda_n < \infty, \tag{5.0.3}$$

and the perturbed system

$$\dot{y} = [A(t) + Q(t)]y, \tag{5.0.4}$$

$$Q \in C(\mathbb{R}_+), \qquad \sup_{t\in\mathbb{R}_+} \|Q(t)\| \leqslant \delta, \tag{5.0.5}$$

with spectrum

$$-\infty < \lambda_1' \leqslant \lambda_2' \leqslant \cdots \leqslant \lambda_n' < \infty. \tag{5.0.6}$$

Under the influence of the perturbations $Q(t)$, the characteristic exponents of system (5.0.1) vary, generally speaking, discontinuously; sometimes a finite shift of the characteristic exponents of the initial system corresponds to an arbitrarily small $\delta$. In this chapter we shall introduce the notion of central exponents of linear systems and shall see that they determine jumps of the exponents under small perturbations. We shall study the properties of the spectrum (5.0.3) itself, its relations with the central exponents, and the influence of these objects on how the exponents vary when passing from system (5.0.1) to system (5.0.4).

## §1. Central exponents

In the investigation of stability in linear approximation, of primary interest is the increase of the greatest and the decrease of the smallest exponent under the influence of perturbations. The key method to solve such problems is provided by upper and lower functions and central exponents of system (5.0.1). These notions have been introduced and studied by Vinograd. In the presentation of the results we follow the monograph **[9]**.

DEFINITION 5.1.1. The greatest exponent $\lambda_n$ of system (5.0.1) is said to be *rigid upwards* if for any $\varepsilon > 0$ there exists a $\delta > 0$ such that $\lambda'_n \leqslant \lambda_n + \varepsilon$, where $\lambda'_n$ is the greatest exponent of system (5.0.4). In the opposite case, $\lambda_n$ is said to be *mobile upwards*, which means that a positive jump of $\lambda_n$ upwards can correspond to an arbitrarily small $\delta$. Analogous motions of *rigidity* and *mobility downwards* can be introduced for the smallest exponent $\lambda_1$.

The bounds of mobility of the exponents are, naturally, of interest.

DEFINITION 5.1.2. Bounded measurable on $\mathbb{R}_+$ functions $r(t)$ and $R(t)$ are said to be *lower* and *upper functions* for system (5.0.1), respectively, if for any solution $x(t)$ of this system the following estimates hold for any $\varepsilon > 0$:

$$d_{r,\varepsilon} \exp\left(\int_s^t (r(\tau) - \varepsilon)\, d\tau\right) \leqslant \frac{\|x(t)\|}{\|x(s)\|} \leqslant D_{R,\varepsilon} \exp\left(\int_s^t (R(\tau) + \varepsilon)\, d\tau\right), \tag{5.1.1}$$

where $t \geqslant s \geqslant 0$ and the quantities $d_{r,\varepsilon}$, $D_{R,\varepsilon}$ do not depend on $s$.

It is clear from the definition that these functions bound the growth of the solutions from below and from above, respectively. Sometimes it is convenient to have the inequalities (5.1.1) in terms of a fundamental matrix, namely, the Cauchy matrix

$$X(t,s) = X(t)X^{-1}(s).$$

LEMMA 5.1.1. *For any fixed $t$ and $s$ the following relations are valid for the Cauchy matrix $X(t,s)$ of a linear system*:

$$\|X(t,s)\| = \max_x \frac{\|x(t)\|}{\|x(s)\|}, \tag{5.1.2}$$

$$\frac{1}{\|X^{-1}(t,s)\|} = \min_x \frac{\|x(t)\|}{\|x(s)\|}. \tag{5.1.3}$$

PROOF.

1. $$\|X(t,s)\| = \max_{\|b\|=1} \|X(t,s)b\| = \max_c \frac{\|X(t,s)c\|}{\|c\|}$$
$$= \max_c \frac{\|X(t)X^{-1}(s)c\|}{\|X(s)X^{-1}(s)c\|} = \max_a \frac{\|X(t)a\|}{\|X(s)a\|} = \max_x \frac{\|x(t)\|}{\|x(s)\|}.$$

2. $$\min_x \frac{\|x(t)\|}{\|x(s)\|} = \frac{1}{\max_x \frac{\|x(s)\|}{\|x(t)\|}} = \frac{1}{\|X(s,t)\|} = \frac{1}{\|X^{-1}(t,s)\|}. \qquad \square$$

The conditions (5.1.1) and (5.1.2) imply that the upper functions $R(t)$ realize the estimate

$$\|X(t,s)\| \leqslant D_{R,\varepsilon} \exp \int_s^t (R(\tau)+\varepsilon)\, d\tau. \tag{5.1.4}$$

DEFINITION 5.1.3. The number $\Omega$ defined as

$$\Omega = \inf_R \left\{ \overline{\lim_{t\to\infty}} \frac{1}{t} \int_0^t R(\tau)\, d\tau \right\} \tag{5.1.5}$$

is called the *upper central exponent* of system (5.0.1). Here the infimum is taken over the set of all the upper functions $R$ of system (5.0.1).

DEFINITION 5.1.4. The number $\omega$ defined as

$$\omega = \sup_r \left\{ \underline{\lim_{t\to\infty}} \frac{1}{t} \int_0^t r(\tau)\, d\tau \right\} \tag{5.1.6}$$

is called the *lower central exponent* of system (5.0.1). In the formula (5.1.6), the supremum is taken over the set of all the lower functions $r$ of system (5.0.1).

REMARK 5.1.1. If in the sets of upper and lower functions we distinguish the subsets of constants, and if we take infimum and supremum over these subsets in (5.1.5) and (5.1.6), then we obtain the definitions of the *upper singular exponent* $\Omega_0$ and the *lower singular exponent* $\omega_0$. The inequalities

$$-M \leqslant \omega_0 \leqslant \omega \leqslant \lambda_1 \leqslant \lambda_n \leqslant \Omega \leqslant \Omega_0 \leqslant M \tag{5.1.7}$$

are obvious. The first and the last inequalities in (5.1.7) follow from Theorem 2.3.1.

REMARK 5.1.2. Lower functions and the lower central exponent do not require special consideration since this problem can be reduced to the investigation of upper functions and the upper central exponent for the adjoint system

$$\dot z = -A^*(t)z. \tag{5.1.8}$$

Indeed,

$$Z(t,s) = [X^{-1}(t,s)]^*.$$

The relations (5.1.3) and (5.1.1) imply the estimate

$$\|X^{-1}(t,s)\| \leqslant D_{r\varepsilon} e^{\int_s^t (-r(\tau)+\varepsilon)\, d\tau}, \tag{5.1.9}$$

and the latter implies our statement.

Thus, the set $\{-r(t)\}$ is the set of upper functions and the set $\{-R(t)\}$ is the set of lower functions for the adjoint system (5.1.8). The following theorem is valid.

THEOREM 5.1.1. *The lower central exponent $\omega$ of system* (5.0.1) *is equal to the upper central exponent of the adjoint system, taken with the opposite sign, and vice versa.*

LEMMA 5.1.2. *Central exponents are invariant under Lyapunov transformations.*

PROOF. Let $x = L(t)w$ be a Lyapunov transformation of system (5.0.1) to the system

$$\dot{w} = (L^{-1}AL - L^{-1}\dot{L})w.$$

The Cauchy matrices of these systems are connected by

$$X(t,s) = X(t)X^{-1}(s) = L(t)W(t)W^{-1}(s)L^{-1}(s),$$

and, by virtue of

$$\|L(t)\| \leqslant K, \qquad \|L^{-1}(t)\| \leqslant K,$$

we have

$$\|X(t,s)\| \leqslant \|W(t,s)\|K^2.$$

For the inverse transformation $w = L^{-1}(t)x$, in a similar way we obtain

$$\|W(t,s)\| \leqslant K^2\|X(t,s)\|.$$

This implies that only the constants are changed in the estimates (5.1.4) and (5.1.9).□

By means of simple examples, we illustrate one of the methods of constructing upper functions and of determining the upper central exponents.

EXAMPLE 5.1.1. Consider a real diagonal system

$$\dot{x} = \operatorname{diag}[a_1(t), \ldots, a_n(t)]x, \qquad a_i \in C(\mathbb{R}_+);$$

its Cauchy matrix has the form

$$X(t,s) = \operatorname{diag}\left[\exp\left(\int_s^t a_1(\tau)\,d\tau\right), \ldots, \exp\left(\int_s^t a_n(\tau)\,d\tau\right)\right].$$

Obviously, any measurable and bounded function $R(t)$ on $\mathbb{R}_+$ is an upper function if

$$\int_s^t a_i(\tau)\,d\tau \leqslant D_{R,\varepsilon} + \int_s^t (R(\tau)+\varepsilon)\,d\tau, \qquad i = 1, \ldots, n, \tag{5.1.10}$$

where $t \geqslant s \geqslant 0$, $\varepsilon > 0$. Here $D_{R,\varepsilon}$ is written instead of $\ln D_{R,\varepsilon}$; this makes no essential difference.

Let us take $T > 0$ and divide the half-axis $\mathbb{R}_+$ by the points $0, T, 2T, \ldots$ into the intervals

$$J_k = [(k-1)T, kT]$$

of length $T$. On each of these intervals we define a function $R^T(t)$, coinciding with that of the functions $a_1(t), \ldots, a_n(t)$ whose integral over this interval is the greatest (or with one of them if there is more than one such function), i.e.,

$$\int_{J_k} R^T(\tau)\,d\tau = \max_i \int_{J_k} a_i(\tau)\,d\tau, \qquad i = 1, \ldots, n, \quad k = 1, 2, \ldots. \tag{5.1.11}$$

Thus, we define a bounded piecewise continuous function $R^T(t)$ on $\mathbb{R}_+$. At the endpoints of the intervals we may set

$$R^T(mT) = 0, \qquad m = 0, 1, \ldots.$$

Let us show that $R^T(t)$ is an upper function. Let the interval $[s, t]$ consist of the integer number $n - m + 1$ of intervals $J_k$; then

$$\int_s^t a_i(\tau)\,d\tau = \sum_{k=m}^{n} \int_{J_k} a_i(\tau)\,d\tau \leqslant \sum_{k=m}^{n} \max_i \int_{J_k} a_{j_\cdot}(\tau)\,d\tau = \int_s^t R^T(\tau)\,d\tau.$$

If the interval $[s, t]$ is arbitrary, then at its ends there appear intervals of length less than $T$. Denote them by $[s, mT]$ and $[nT, t]$, assuming that

$$[mT, nT] \subset [s, t],$$

and

$$(m-1)T < s, \qquad (n+1)T > t.$$

Then we have

$$\begin{aligned}\int_s^t a_i(\tau)\,d\tau &= \int_s^{mT} a_i\,d\tau + \int_{mT}^{nT} a_i\,d\tau + \int_{nT}^t a_i\,d\tau \\ &\leqslant \int_s^{mT} R^T\,d\tau + \int_{mT}^{nT} R^T\,d\tau + \int_{nT}^t R^T\,d\tau \\ &\quad - \int_s^{mT} (R^T - a_i)\,d\tau - \int_{nT}^t (R^T - a_i)\,d\tau.\end{aligned}$$

Since

$$|R^T(t) - a_i(t)| \leqslant 2M,$$

and the lengths of the last two intervals of integration do not exceed $T$, we finally obtain

$$\int_s^t a_i(\tau)\,d\tau \leqslant 4MT + \int_s^t R^T(\tau)\,d\tau, \qquad i = 1, \dots, n.$$

By virtue of (5.1.10), this implies that $R^T(t)$ is an upper function; hence,

$$\overline{\lim_{t\to\infty}} \frac{1}{t}\int_0^t R^T(\tau)\,d\tau \geqslant \Omega. \tag{5.1.12}$$

Let us show that the limit in the inequality (5.1.12) is realized along the sequence

$$t_n = nT, \qquad n = 1, 2, \dots.$$

Indeed, for

$$nT \leqslant t \leqslant (n+1)T$$

we have

$$\begin{aligned}&\left|\frac{1}{t}\int_0^t R^T\,d\tau - \frac{1}{nT}\int_0^{nT} R^T\,d\tau\right| \\ &\quad\leqslant \left|\frac{1}{t}\int_0^t R^T\,d\tau - \frac{1}{nT}\int_0^t R^T\,d\tau\right| + \left|\frac{1}{nT}\int_0^t R^T\,d\tau - \frac{1}{nT}\int_0^{nT} R^T\,d\tau\right| \\ &\quad\leqslant \frac{t-nT}{nT}\left|\frac{1}{t}\int_0^t R^T\,d\tau\right| + \left|\frac{1}{nT}\int_{nT}^t R^T\,d\tau\right| \\ &\quad\leqslant \frac{t-nT}{nT}M + \frac{t-nT}{nT}M \leqslant \frac{2M}{n} \underset{n\to\infty}{\longrightarrow} 0,\end{aligned}$$

since $t - nT \leqslant T$. Therefore, from the inequality (5.1.11) we obtain

$$\Omega_T = \lim_{n\to\infty} \frac{1}{nT}\int_0^{nT} R^T(\tau)\,d\tau \geqslant \Omega, \tag{5.1.13}$$

or

$$\bar{\Omega} = \inf_{T>0} \Omega_T \geqslant \Omega.$$

Let us prove the reverse inequality, which would imply that $\bar{\Omega} = \Omega$. Take an $\varepsilon > 0$ and an upper function $R(t)$, defined in some other way such that

$$\overline{\lim_{t\to\infty}} \frac{1}{t} \int_0^t R(\tau)\, d\tau \leqslant \Omega + \varepsilon,$$

$$\max_i \int_{J_k} a_i(\tau)\, d\tau \leqslant D_{R,\varepsilon} + \int_{J_k} (R(\tau) + \varepsilon)\, d\tau.$$

On the interval $[0, nT]$ we have

$$\int_0^{nT} R^T(\tau)\, d\tau \leqslant nD_{R,\varepsilon} + \int_0^{nT} (R(\tau) + \varepsilon)\, d\tau,$$

or

$$\frac{1}{nT} \int_0^{nT} R^T(\tau)\, d\tau \leqslant \frac{D_{R,\varepsilon}}{T} + \varepsilon + \frac{1}{nT} \int_0^{nT} R(\tau)\, d\tau.$$

Letting $n \to \infty$, we obtain

$$\Omega_T \leqslant D_{R,\varepsilon}/T + \Omega + 2\varepsilon.$$

Therefore,

$$\bar{\Omega} = \inf_{T>0} \Omega_T \leqslant \Omega;$$

thus, we have $\bar{\Omega} = \Omega$, i.e., the upper function $R^T(t)$ constructed above realizes the upper central exponent.

EXAMPLE 5.1.2. Consider the case of an autonomous system

$$\dot{x} = Ax, \qquad x \in \mathbb{C}^n, \quad A = \text{const}.$$

Let $\alpha_1, \alpha_2, \ldots, \alpha_n$ be the eigenvalues of the matrix $A$ and

$$\Lambda = \max_i \{\operatorname{Re} \alpha_i\}, \qquad \lambda = \min_i \{\operatorname{Re} \lambda_i\}.$$

From the estimate (1.3.10) we have

$$\|X(t,s)\| = \|\exp A(t-s)\| \leqslant D_\varepsilon \exp(\Lambda + \varepsilon)(t-s), \qquad t \geqslant s \geqslant 0.$$

It follows that $R(t) = \Lambda$; hence, $\Lambda = \Omega = \Omega_0$. Estimating the Cauchy matrix from below, or using the adjoint system, we obtain $\lambda = \omega_0 = \omega$, since $r(t) = \lambda$.

EXAMPLE 5.1.3. Consider the system

$$\dot{x}_1 = x_1, \qquad \dot{x}_2 = (2 + \cos t)x_2.$$

Here

$$X(t,s) = \operatorname{diag}[e^{t-s}, e^{2(t-s)+\sin t - \sin s}] = [x_1(t,s), x_2(t,s)],$$

and

$$\chi[x_1] = 1, \qquad \chi[x_2] = 2.$$

For all $t \in \mathbb{R}$ we have

$$a_1(t) = 1 \leqslant a_2 = 2 + \cos t;$$

therefore, we can take $R(t) = 2 + \cos t$. Hence,

$$\Omega = \overline{\lim_{t\to\infty}} \frac{1}{t} \int_0^t R(\tau)\, d\tau = 2,$$

i.e., $\Omega = \lambda_2$. We can take $r(t) = 1$ as a lower function; hence, $\omega = \lambda_1 = 1$.

EXAMPLE 5.1.4. Consider the system

$$\dot{x}_1 = 0, \qquad \dot{x}_2 = \pi \sin \pi\sqrt{t}\, x_2.$$

Then

$$X(t,s) = (x_1, x_2) = \begin{pmatrix} 1 & 0 \\ 0 & e^{p(t)-p(s)} \end{pmatrix},$$

where

$$p(t) = \int_0^t \pi \sin \pi\sqrt{\tau}\, d\tau = \frac{2}{\pi}(\sin \pi\sqrt{t} - \pi\sqrt{t}\cos \pi\sqrt{t}).$$

In our case,

$$\lambda_1 = 0, \qquad \lambda_2 = \overline{\lim_{t\to\infty}} \frac{1}{t} p(t) = 0.$$

As an upper function we take

$$R(t) = (\pi \sin \pi\sqrt{t} + |\pi \sin \pi\sqrt{t}|)/2.$$

Note that

$$R(t) > 0 \qquad \text{for} \qquad t \in (4k^2, (2k+1)^2)$$

and

$$R(t) \equiv 0 \qquad \text{for} \qquad t \in [(2k+1)^2, 4(k+1)^2], \qquad k = 0, 1, \dots.$$

Let us show that

$$\overline{\lim_{t\to\infty}} \frac{1}{t} \int_0^t R(\tau)\, d\tau$$

can be realized along the sequence $t_k = k^2$. Let

$$(2k)^2 \leqslant t \leqslant (2k+1)^2.$$

By analogy with Example 5.1.1, we have

$$\left| \frac{1}{t} \int_0^t R(\tau)\, d\tau - \frac{1}{4k^2} \int_0^{4k^2} R(t)\, dt \right| \underset{k\to\infty}{\longrightarrow} 0.$$

By straightforward calculations, we obtain

$$\int_{(2k)^2}^{(2k+1)^2} R(\tau)\, d\tau = \frac{2}{\pi}(\sin \pi\sqrt{t} - \pi\sqrt{t}\cos \pi\sqrt{t})\Big|_{(2k)^2}^{(2k+1)^2} = 2(2k+1+2k) = 2(4k+1),$$

or

$$\int_0^{(2n+1)^2} R(t)\, dt = \sum_{k=0}^{n} \int_{(2k)^2}^{(2k+1)^2} R(t)\, dt = \sum_{k=0}^{n} 2(4k+1) = 2(2n+1)(n+1).$$

Finally, we have

$$\Omega = \lim_{n\to\infty} \frac{1}{(2n+1)^2} \int_0^{(2n+1)^2} R(t)\, dt = \lim_{n\to\infty} \frac{2(2n+1)(n+1)}{(2n+1)^2} = 1.$$

Taking

$$r(t) = (\pi \sin \pi\sqrt{t} - |\pi \sin \pi\sqrt{t}|)/2,$$

we obtain $\omega = -1$.

Let us turn again to systems (5.0.1) and (5.0.4), and let $\|Q(t)\| \to 0$, $t \to \infty$. Let us find out what happens to upper functions under this condition.

THEOREM 5.1.2. *Perturbations of a linear system that tend to zero as $t \to \infty$ preserve the sets of upper and lower functions and do not change the central exponents.*

PROOF. Thus, we have

$$\|Q(t)\| \leqslant \delta(t) \underset{t\to\infty}{\longrightarrow} 0. \tag{5.1.14}$$

Let us write the solution $y(t)$ of system (5.0.4) using the method of variation of parameters,

$$y(t) = X(t,0)y(0) + \int_0^t X(t,s)Q(s)y(s)\,ds,$$

or

$$\|y(t)\| \leqslant \|X(t,0)\|\|y(0)\| + \int_0^t \|X(t,s)\|\|Q(s)\|\|y(s)\|\,ds.$$

Take an $\varepsilon > 0$ and one of the upper functions $R(t)$ of system (5.0.1). By virtue of the estimate (5.1.4), we have

$$\|y(t)\| \leqslant \|y(0)\|D_{R,\varepsilon}e^{\int_0^t (R+\varepsilon/2)\,d\tau} + \int_0^t D_{R,\varepsilon}e^{\int_s^t (R+\varepsilon/2)\,d\tau}\delta(s)\|y(s)\|\,ds,$$

or

$$\|y(t)\|e^{-\int_0^t (R+\varepsilon/2)\,d\tau} \leqslant \|y(0)\|D_{R,\varepsilon} + \int_0^t D_{R,\varepsilon}\delta(s)e^{-\int_0^s (R+\varepsilon/2)\,d\tau}\|y(s)\|\,ds.$$

By the Gronwall-Bellman lemma, we have

$$\|y(t)\|e^{-\int_0^t (R+\varepsilon/2)\,d\tau} \leqslant \|y(0)\|D_{R,\varepsilon}e^{\int_0^t D_{r,\varepsilon}\delta(s)\,ds}. \tag{5.1.15}$$

By the condition (5.1.14), there exists a $T > 0$ such that

$$\delta(t) < \varepsilon/(2D_{R,\varepsilon}) \qquad \text{for} \qquad t \geqslant T.$$

We continue the estimate (5.1.15):

$$\begin{aligned}\|y(t)\| &\leqslant D_{R,\varepsilon}\|y(0)\| \exp\left[\int_0^t (R+\varepsilon/2)\,d\tau + \int_0^T D_{R,\varepsilon}\delta(s)\,ds + \int_T^t D_{R,\varepsilon}\delta(s)\,ds\right] \\ &\leqslant D_{R,\varepsilon}\|y(0)\|e^{D_{R,\varepsilon}T\delta} \exp\int_0^t (R(s)+\varepsilon)\,ds.\end{aligned}$$

Here $\delta = \max_{[0,T]}\delta(t)$. It follows from the inequality obtained that $R(t)$ is an upper function for system (5.0.4) as well. Changing the roles of systems (5.0.1) and (5.0.4) in the above arguments, we come to the conclusion that an upper function for system (5.0.4) is also an upper function for system (5.0.1). Thus, the sets of upper functions of these systems coincide; hence, the upper central exponents also coincide. For lower functions and lower central exponents, the arguments are carried out by passing to the adjoint systems. □

Let us return to the question of the mobility of the bounds of the extreme exponents $\lambda_1$ and $\lambda_n$ of system (5.0.1) under small perturbations, i.e., under the passage to system (5.0.4).

DEFINITION 5.1.5. A number $\Omega'$ is called an *upper bound of mobility of* $\lambda_n$, if for any $\varepsilon > 0$ there exists a $\delta > 0$ such that

$$\lambda'_n \leqslant \Omega' + \varepsilon$$

($\lambda'_n$ belongs to the spectrum (5.0.6)).

Obviously, $\lambda_n \leqslant \Omega'$ (consider zero perturbations); therefore, $\lambda_n \leqslant \inf \Omega'$. We give a theorem that makes the value of $\Omega'$ more precise.

THEOREM 5.1.3 (Vinograd). *For any $\varepsilon > 0$ there exists a $\delta > 0$ such that the greatest exponent of system* (5.0.4) *does not exceed the quantity $\Omega + \varepsilon$.*

PROOF. Repeat the proof of Theorem 5.1.2, omitting (5.1.14) up to the inequality (5.1.15) inclusive, just taking into account that $\delta = \text{const}$. Then we write

$$\|y(t)\| \leqslant \|y(0)\| D_{R,\varepsilon} \exp\left(\int_0^t [R(\tau) + \varepsilon_1/2 + D_{R,\varepsilon}\delta]\, d\tau\right). \tag{5.1.16}$$

Here $\varepsilon_1 = \varepsilon/2$, the function $R(t)$ is assumed to be such that

$$\overline{\lim_{t\to\infty}} \frac{1}{t} \int_0^t R(\tau)\, d\tau \leqslant \Omega + \frac{\varepsilon}{2}, \tag{5.1.17}$$

and we choose $\delta$ equal to $\varepsilon_1/(2D_{R,\varepsilon})$.

From the inequality (5.1.16), taking into account (5.1.17), we have

$$\chi[y] \leqslant \Omega + \varepsilon/2 + \varepsilon_1/2 + \varepsilon_1/2 = \Omega + \varepsilon.$$

Thus, given $\varepsilon > 0$ we have defined $R(t)$ so that the inequality (5.1.17) is satisfied, then we have determined $D_{R,\varepsilon}$, and, finally, $\delta$. □

COROLLARY 5.1.1. *The upper central exponent $\Omega$ bounds from above the mobility of the greatest exponent of the system.*

Analogously, we can introduce a lower bound of mobility of the smallest exponent $\lambda_1$ of system (5.0.1), and by passing to adjoint systems we can show that this bound is realized by the lower central exponent $\omega$.

Now let us return to the greatest exponent. Its mobility upwards is bounded by the number $\Omega$, but this is an estimate from above. Is it exceedingly crude? Can arbitrarily small perturbations lead to a "jump" of the greatest exponent $\lambda_n$ into a neighborhood of $\Omega$?

DEFINITION 5.1.6. A number $\bar{\Omega}$ is said to be the *attainable upper bound of mobility* of the greatest exponent $\lambda_n$ if it is an upper bound, and for any $\varepsilon > 0$ there exist systems (5.0.4) with an arbitrarily small $\delta$ and the greatest exponent $\lambda'_n > \bar{\Omega} - \varepsilon$.

Millionshchikov proved that central exponents are attainable **[29]**. We omit the proof of the following theorem.

THEOREM 5.1.4 (Millionshchikov). *The central exponents of linear systems are attainable, i.e., for any $\varepsilon > 0$ there exist piecewise continuous matrices $B_\varepsilon(t)$ and $C_\varepsilon(t)$,*

$$\|B_\varepsilon(t)\| \leqslant \varepsilon, \qquad \|C_\varepsilon(t)\| \leqslant \varepsilon \qquad \textit{for} \qquad t \in \mathbb{R}_+,$$

*such that the greatest exponent of the system*

$$\dot{y} = [A(t) + B_\varepsilon(t)]y$$

*is no less than* $\Omega - \varepsilon$, *and the smallest exponent of the system*

$$\dot{y} = [A(t) + C_\varepsilon(t)]y$$

*is no greater than* $\omega + \varepsilon$.

COROLLARY 5.1.2. *The greatest exponent is rigid upwards if and only if it coincides with the upper central exponent.*

The corollary follows from Definition 5.1.1 and Theorems 5.1.3 and 5.1.4.

Following Vinograd **[9]**, we show that the upper central exponent for a diagonal system is attainable.

THEOREM 5.1.5 (Vinograd). *Take a real system*

$$\dot{x} = \operatorname{diag}[a_1(t), a_2(t), \dots, a_n(t)]x, \qquad a_i \in C(\mathbb{R}), \tag{5.1.18}$$

*and let* $\Omega$ *be its upper central exponent.*

*Then the perturbed system*

$$\begin{aligned} \dot{x}_1 &= a_1(t)x_1 + && && +\delta x_n,\\ \dot{x}_2 &= \delta x_1 + a_2(t)x_2, && &&\\ \dot{x}_3 &= && \delta x_2 + a_3(t)x_3, &&\\ &\cdots\cdots\cdots && \cdots\cdots\cdots &&\\ \dot{x}_n &= && && \delta x_{n-1} + a_n(t)x_n, \end{aligned} \tag{5.1.19}$$

*where* $\delta$ *is arbitrarily small, has the characteristic exponent* $\lambda'_n \geqslant \Omega$.

PROOF. To simplify calculations, we write system (5.1.19) in the form

$$\dot{x}_\alpha = a_\alpha(t)x_\alpha + \delta x_{\alpha-1}, \tag{5.1.20}$$

where the index $\alpha$ is assumed to be cyclic $\bmod n$. By means of the Picard method **[5]**, we construct the solution of this system with the initial condition

$$x_\alpha(0) = 1, \qquad \alpha = 1, \dots, n.$$

We pass from system (5.1.20) to the integral equations

$$x_\alpha = \exp\left(\int_0^t a_\alpha(\tau)\,d\tau\right) + \delta \int_0^t x_{\alpha-1}(\tau)\left[\exp\left(\int_\tau^t a_\alpha(\xi)\,d\xi\right)\right]d\tau,$$

and establish the form of the Picard approximations,

$$x_\alpha^{(k)}(t) = \exp\left(\int_0^t a_\alpha(\tau)\,d\tau\right) + \delta \int_0^t x_{\alpha-1}^{(k-1)}(\tau)\left[\exp\left(\int_\tau^t a_\alpha(\xi)\,d\xi\right)\right]d\tau. \tag{5.1.21}$$

Hence, we have

$$x_\alpha^{(1)}(t) = x_\alpha^{(0)}(t) + \delta \int_0^t \left(\exp\left[\int_0^{t_1} a_{\alpha-1}\,d\xi + \int_{t_1}^t a_\alpha d\xi\right] dt_1\right)$$

(here we have set $\tau = t_1$),

$$x_\alpha^{(2)}(t) = x_\alpha^{(1)}(t) + \delta^2 \int_0^t \int_0^{t_2} \left(\exp\left[\int_0^{t_1} a_{\alpha-2}\,d\xi + \int_{t_1}^{t_2} a_{\alpha-1}\,d\xi + \int_{t_2}^t a_\alpha\,d\xi\right]\right) dt_1\,dt_2$$

(here $\tau = t_2$). Further, we proceed by induction. Let

$$x_\alpha^{(k)}(t) = x_\alpha^{(k-1)}(t) + \delta^k J_\alpha^{(k)}(t), \tag{5.1.22}$$

where

$$J_\alpha^{(k)}(t) = \int_0^t \int_0^{t_k} \cdots \int_0^{t_2} \left( \exp\left[ \int_0^{t_1} a_{\alpha-k}\, d\xi + \int_{t_1}^{t_2} a_{\alpha-k+1}\, d\xi \right.\right. \\ \left.\left. + \cdots + \int_{t_k}^{t} a_\alpha\, d\xi \right] \right) dt_1 \cdots dt_k. \tag{5.1.23}$$

We change $k$ to $k+1$ in (5.1.21) and, using (5.1.22), we write

$$\begin{aligned} x_\alpha^{(k+1)}(t) &= x_\alpha^{(k)}(t) + \delta \int_0^t \delta^k J_{\alpha-1}^{(k)}(\tau) \left( \exp \int_\tau^t a_\alpha\, d\xi \right) d\tau \\ &= x_\alpha^{(k)}(t) + \delta^{k+1} \int_0^t \left( \exp \int_\tau^t a_\alpha\, d\xi \right) \int_0^\tau \int_0^{t_k} \cdots \int_0^{t_2} \\ &\quad \times \left( \exp\left[ \int_0^{t_1} a_{\alpha-k}\, d\xi + \int_{t_1}^{t_2} a_{\alpha-k+1}\, d\xi + \cdots + \int_{t_k}^{\tau} a_{\alpha-1}\, d\xi \right] \right) \\ &\quad \times dt_1\, dt_2 \cdots dt_k\, d\tau. \end{aligned}$$

Let us insert $\exp \int_\tau^t a_\alpha\, d\xi$ under the sign of the $k$-tuple integral; then, setting $\tau = t_{k+1}$, we obtain

$$x_\alpha^{(k+1)}(t) = x_\alpha^{(k)}(t) + \delta^{k+1} J_\alpha^{(k+1)}(t);$$

this proves that the equality (5.1.22) holds and that it can be written in the form

$$x_\alpha^{(k)} = \exp \int_0^t a_\alpha\, d\xi + \sum_{r=1}^{k} \delta^r J_\alpha^{(r)}(t).$$

Since the Picard approximations converge for all $t \in \mathbb{R}$, we have the solution required,

$$x_\alpha(t) = \exp \int_0^t a_\alpha\, d\xi + \sum_{k=1}^{\infty} \delta^k J_\alpha^{(k)}(t), \qquad \alpha = 1, \ldots, n. \tag{5.1.24}$$

Now we pass to the proof of the statement that the solution obtained has the characteristic exponent $\chi[x] \geqslant \Omega$. The terms of the series (5.1.24) are positive; therefore,

$$\|x(t)\| = \left( \sum_{k=1}^{n} |x_\alpha(t)|^2 \right)^{1/2} \geqslant x_n(t) > \delta^k J_n^{(k)}(t); \tag{5.1.25}$$

the choice of the index is not important but it is convenient to specify it. Let us show that we can define a sequence $t_m \to \infty$ so that the corresponding sequence $J_n^{(k(m))}(t)$ has maxima at the points $t_m$, the values of $J_n^{(k(m))}(t_m)$ growing sufficiently rapidly with $m$. We shall compare $\|x(t)\|$ with the function $J_n^{(k(m))}(t)$ at the points $t_m$; this allows us to estimate from below the exponent of $\|x(t)\|$ and verify that $\chi[x] \geqslant \Omega$.

We pass to explicit calculations. Take an arbitrarily large $T > n$, where $n$ is the order of the system, and consider the function $J_n^{(k)}(t)$ for

$$k = nm - 1, \qquad t = mT, \qquad m \in \mathbb{N}.$$

Write out the sum of integrals entering the exponent of the exponential in the integrand and, taking into account that for the indicated value of $k$ the index has run over $m$ cycles, we combine these integrals in $m$ groups:

$$
\begin{aligned}
z = {} & \underbrace{\int_0^{t_1} a_1\, d\xi + \int_{t_1}^{t_2} a_2\, d\xi + \cdots + \int_{t_{n-1}}^{t_n} a_n\, d\xi}_{\text{1st group}} \\
& + \underbrace{\int_{t_n}^{t_{n+1}} a_1\, d\xi + \int_{t_{n+1}}^{t_{n+2}} a_2\, d\xi + \cdots + \int_{t_{2n-1}}^{t_{2n}} a_n\, d\xi}_{\text{2nd group}} \\
& + \cdots + \underbrace{\int_{t_{n(m-1)}}^{t_{n(m-1)+1}} a_1\, d\xi + \cdots + \int_{t_{nm-1}}^{t_{nm}} a_n\, d\xi}_{m\text{th group}} .
\end{aligned}
\tag{5.1.26}
$$

The interval $[0, mT]$ is divided by the points $T, 2T, \dots, (m-1)T$ into the intervals $\Delta_1, \Delta_2, \dots, \Delta_m$ of length $T$. Let us turn to the method of constructing the upper function (described in Example 5.1.1) such that $R^T(t)$ can be assumed to coincide with one of the functions $a_{i_k}(t)$ on the interval $\Delta_k$, i.e.,

$$
R^T(t) = a_{i_k}(t), \qquad t \in \Delta_k .
$$

This implies

$$
\int_0^{mT} R^T(\tau)\, d\tau = \int_0^T a_{i_1}\, d\tau + \int_T^{2T} a_{i_2}\, d\tau + \cdots + \int_{(m-1)T}^{mT} a_{i_m}(\tau)\, d\tau. \tag{5.1.27}
$$

We show that the quantity (5.1.27) is a particular value of $z$ for some specific values of the variables of integration. Let us associate to each interval $\Delta_k$ the $k$th group from the collection (5.1.26) and study this problem for $k = 1$ (the arguments for $k > 1$ are similar). The first summand in (5.1.27) is obtained if, in the first group, the integral $\int_{t_{i_1-1}}^{t_{i_1}} a_{i_1}(\xi)\, d\xi$ is extended to the whole interval $\Delta_1$, i.e., the points preceding $t_{i_1-1}$ are contracted to zero and the succeeding ones, to $t_n = T$; this gives

$$
\begin{gathered}
0 = t_1 = \cdots = t_{i_1-1}, \\
t_{i_1} = t_{i_1+1} = \cdots = t_n = T.
\end{gathered}
\tag{5.1.28}
$$

Further, determine the subsequent deviations of the variables on the interval $\Delta_1$ from their values in (5.1.28). On $\Delta_1 = [0, T]$, we place $n$ intervals of unit length grouping them as follows: $i_1 - 1$ ones are adjacent to zero, and the others, to $T$. Now let each $t_i$ run over only the corresponding interval indicated in Figure 2.

Note that the deviation of $t_i$ from its value (5.1.28) does not exceed $n$. Note also that the variation of each variable $t_1, t_2, \dots, t_{n-1}$ affects two integrals (the lower limit of one of them and the upper limit of the other), and $t_n$ affects only the last integral; moreover,

$$
|a_i(t)| \leqslant M, \qquad i = 1, \dots, n.
$$

Therefore, the sum of the integrals of the first group under the deviation described does not exceed $2M(n-1)$, i.e.,

$$
\text{1st group} \ \geqslant \int_0^T R^T(\tau)\, d\tau - 2Mn(n-1). \tag{5.1.29}
$$

FIGURE 2

Repeating similar arguments for $\Delta_2, \dots, \Delta_m$, from formula (5.1.27) and the inequality (5.1.29) we obtain

$$z \geqslant \int_0^{mT} R^T(\tau)\,d\tau - 2Mn(mn-1) \geqslant \int_0^{mT} R^T(\tau)\,d\tau - 2Mn^2m. \tag{5.1.30}$$

We return to the integral $J_\alpha^{(k)}(t)$ for $k = nm - 1$, $t = mT$. The integrand is positive; therefore, when the domain diminishes, the integral decreases. According to the definition (5.1.23), its domain of integration is the interval

$$0 \leqslant t_1 \leqslant t_2 \leqslant \cdots \leqslant t_{nm-1} \leqslant t = t_{nm},$$

and the variables $t_i$ run over an $(mn-1)$-dimensional cube with edge of unit length, i.e., a subdomain of unit volume. Therefore,

$$J_n^{(nm-1)}(mT) \geqslant \exp\left[\int_0^{mT} R^T(\tau)\,d\tau - 2Mn^2m\right].$$

We return to the estimate (5.1.25). It is of interest to us for sufficiently large $t$. Let

$$mT \leqslant t \leqslant (m+1)T.$$

Then

$$\begin{aligned}
\|x(t)\| &\geqslant \delta^{nm-1} J_n^{(nm-1)}(t) \\
&\geqslant \delta^{nm-1} J_n^{(nm-1)}(mT) \\
&\geqslant \delta^{mn-1} \exp\left[\int_0^{mT} R^T\,d\tau - 2Mn^2m\right] \\
&= \exp\left[\int_0^{mT} R^T\,d\tau - 2Mn^2m + (nm-1)\ln\delta\right].
\end{aligned}$$

By inequality (5.1.13), for any $\varepsilon > 0$ there exists an $N$ such that for any $m \geqslant N$ we have

$$\int_0^{mT} R^T(\tau)\,d\tau \geqslant (\Omega - \varepsilon)mT. \tag{5.1.31}$$

Hence,

$$\begin{aligned}
\chi[x] &= \overline{\lim_{t\to\infty}} \frac{1}{t} \ln\|x(t)\| \\
&\geqslant \overline{\lim_{m\to\infty}} \frac{1}{mT} \ln\|x(mT)\| \overset{(5.1.31)}{\geqslant} \lim_{m\to\infty} \frac{(nm-1)\ln\delta - 2Mn^2m + (\Omega-\varepsilon)mT}{mT} \\
&= \Omega - \varepsilon - \frac{2Mn^2 - n\ln\delta}{T} = \Omega - \varepsilon - \varepsilon_1.
\end{aligned}$$

Note that $\varepsilon_1 > 0$, since $\delta$ is small, and $\varepsilon_1 \to 0$ as $T \to \infty$. Thus, we have shown that system (5.1.19) has a characteristic exponent which is not less than $\Omega$. □

COROLLARY 5.1.3. *In diagonal systems the upper central exponent $\Omega$ is always the least upper bound of mobility of the greatest exponent, and the latter is rigid if and only if it coincides with $\Omega$.*

Let us illustrate the theorem proved on the example of the system considered in Example 5.1.4.

EXAMPLE 5.1.5. The system

$$\begin{aligned} \dot{x}_1 &= 0, \\ \dot{x}_2 &= \pi \sin \pi\sqrt{t}x_2, \end{aligned} \tag{5.1.32}$$

as follows from Example 5.1.4, has $\lambda_1 = \lambda_2 = 0$, $\Omega = 1$. We show that the perturbed system

$$\begin{aligned} \dot{x}_1 &= \delta x_2, \\ \dot{x}_2 &= \delta x_1 + \pi \sin \pi\sqrt{t}x_2 \end{aligned} \tag{5.1.33}$$

has a characteristic exponent $\lambda'$ such that $\lambda' \geqslant \Omega = 1$. We repeat the arguments of Theorem 5.1.5, and consider $J_\alpha^{(r)}(t)$ for $r$ even and $\alpha = 2$,

$$\begin{aligned} J_2^{(r)}(t) = \int_0^t \int_0^{t_r} \cdots \int_0^{t_2} \left( \exp\left[ \int_0^{t_1} a_2\, d\tau + \int_{t_2}^{t_3} a_2\, d\tau + \cdots + \int_{t_r}^{t} a_2\, d\tau \right] \right) \\ \times dt_1\, dt_2 \cdots dt_r. \end{aligned} \tag{5.1.34}$$

Since

$$R(t) = (\pi \sin \pi\sqrt{t} + |\pi \sin \pi\sqrt{t}|)/2,$$

considered in Example 5.1.4, is a sharp upper function, we can modify the partition of $\mathbb{R}_+$, and, instead of the intervals $\Delta_k$ of length $T$, we can take the intervals

$$\Delta_k = [T_{2k}, T_{2k+1}],$$

where

$$T_l = l^2, \qquad k = 0, 1, \dots.$$

Let us estimate $J_2^{(2m)}(T_{2m+1})$. The function

$$z(t_1, t_2, \dots, t_r, t) = \int_0^{t_1} a_2\, d\tau + \int_{t_2}^{t_3} a_2\, d\tau + \cdots + \int_{t_r}^{t} a_2\, d\tau$$

for $t_i = T_i$, $t = T_{2m+1}$ coincides with

$$\int_0^{T_{2m+1}} R(\tau)\, d\tau = \sum_{k=0}^{m} \int_{(2k)^2}^{(2k+1)^2} \pi \sin \pi\sqrt{t}\, dt = 2(2m+1)(m+1),$$

and when $t_i$ changes in the bounds

$$|t_i - T_i| \leqslant 1,$$

$z(t_1, t_2, \dots, t_r, t)$ changes by at most $4m\pi$. Indeed, let us turn to the inequality (5.1.30). In our case $M = \pi$, $n = 2$, $mn - 1 = r = 2m$, the result of the deviation should be divided by 2 because $a_1 \equiv 0$, and there are only two diagonal coefficients. Repeating the subsequent arguments of the theorem, we obtain

$$\begin{aligned}\|x(T_{2m+1})\| &\geqslant \delta^{2m}\exp\left(\int_0^{T_{2m+1}} R(t)\,dt - 4\pi m\right)\\ &= \exp\left(\int_0^{T_{2m+1}} R(t)\,dt - 4\pi m + 2m\ln\delta\right).\end{aligned}$$

Hence,

$$\begin{aligned}\chi[x] &= \overline{\lim_{t\to\infty}}\,\frac{1}{t}\ln\|x(t)\|\\ &\geqslant \lim_{m\to\infty}\frac{1}{T_{2m+1}}\ln\|x(T_{2m+1})\|\\ &\geqslant \lim_{m\to\infty}\frac{2(2m+1)(m+1) - 4\pi m + 2m\ln\delta}{(2m+1)^2} = 1.\end{aligned}$$

EXAMPLE 5.1.6. Consider a variant of perturbation of system (5.1.32). Instead of system (5.1.33), we consider the system

$$\begin{aligned}\dot{x}_1 &= (\delta/\sqrt{t})x_2,\\ \dot{x}_2 &= (\delta/\sqrt{t})x_1 + \pi\sin\pi\sqrt{t}\,x_2,\end{aligned} \tag{5.1.35}$$

where, as above, $\delta$ is small, but the perturbations in this system tend to zero, i.e., the central exponents of systems (5.1.32) and (5.1.35) coincide (Theorem 5.1.2).

We indicate the differences from the arguments of Example 5.1.5. The function $J_2^{(r)}(t)$ under the sign of the integral contains the extra factor $1/(\sqrt{t_1}\sqrt{t_2}\dots\sqrt{t_r})$ in comparison with (5.1.34). For $t = T_{2m+1}$ this factor is estimated from below by the quantity

$$1/(T_{2m+1})^m = 1/(2m+1)^{2m}.$$

Then

$$\begin{aligned}\|x(T_{2m+1})\| &\geqslant \delta^{2m}\exp\left(\int_0^{T_{2m+1}} R(t)\,dt - 4\pi m\right)\frac{1}{(2m+1)^{2m}}\\ &\geqslant \exp\left(\int_0^{T_{2m+1}} R(t)\,dt - 4\pi m + 2m\ln\delta - 2m\ln(2m+1)\right).\end{aligned}$$

Finally, we have

$$\chi[x] \geqslant \lim_{m\to\infty}\frac{2(2m+1)(m+1) - 4\pi m + 2m\ln\delta - 2m\ln(2m+1)}{(2m+1)^2} = 1.$$

In the preceding example we stopped at this inequality. Here we can state the exact equality $\chi[x] = 1$. Why? The upper central exponent of system (5.1.35) is equal to unity; therefore, the inequality $\chi[x] > 1$ is impossible.

## §2. On the stability of characteristic exponents

Let us turn again to systems (5.0.1) and (5.0.4). From the examples of the previous section it is clear that small perturbations of the coefficients of a system can lead to finite shifts of the characteristic exponents. Thus, in Example 5.1.6 the perturbation

$$Q(t) = \begin{pmatrix} 0 & \delta/\sqrt{t} \\ \delta/\sqrt{t} & 0 \end{pmatrix}$$

shifts the greatest exponent $\lambda_n$ of system (5.1.32) by one to the right and, at the same time, for system (5.1.35) the perturbation

$$Q(t) = \begin{pmatrix} 0 & -\delta/\sqrt{t} \\ -\delta/\sqrt{t} & 0 \end{pmatrix}$$

shifts the greatest exponent by one to the left, since it returns us back to system (5.1.32). A similar situation is possible for the interior points of the spectrum (5.0.3) of system (5.0.1).

Nevertheless, a sufficiently ample class of systems (5.0.1) possesses spectra that are rigid under small perturbations. In what follows we shall find which properties of a system determine this behavior, prove necessary and sufficient conditions for the rigidity of the spectrum, and show, e.g., that autonomous, periodic systems, as well as systems that are reducible and almost reducible to autonomous systems, possess such rigidity.

DEFINITION 5.2.1. The characteristic exponents of system (5.0.1) are said to be *stable* if for any $\varepsilon > 0$ there exists a $\delta > 0$ such that the inequality $\sup_{t\in\mathbb{R}_+} \|Q(t)\| < \delta$ implies the inequality

$$|\lambda_i - \lambda_i'| < \varepsilon, \qquad i = 1, \ldots, n. \tag{5.2.1}$$

EXAMPLE 5.2.1. The characteristic exponent of a linear scalar equation is always stable. The equation $\dot{x} = a(t)x$ has the solution

$$x(t) = x(0) \exp \int_0^t a(\tau)\, d\tau.$$

Therefore,

$$\lambda = \overline{\lim_{t\to\infty}} \frac{1}{t} \int_0^t a(\tau)\, d\tau.$$

The perturbed system

$$\dot{y} = [a(t) + q(t)]y, \qquad \text{where} \qquad |q(t)| \leqslant \delta \qquad \text{for} \qquad t \in \mathbb{R}_+,$$

has the solution

$$y(t) = y(0) \exp \int_0^t [a(\tau) + q(\tau)]\, d\tau,$$

or

$$\exp\left(\int_0^t a(\tau)\, d\tau - \delta t\right) \leqslant \frac{y(t)}{y(0)} \leqslant \exp\left(\int_0^t a(\tau)\, d\tau + \delta t\right),$$

and, according to the properties of the exponents, we have

$$\lambda - \delta \leqslant \lambda' \leqslant \lambda + \delta \qquad \Rightarrow \qquad |\lambda - \lambda'| \leqslant \delta.$$

Here

$$\chi[y] = \lambda' = \overline{\lim_{t\to\infty}} \frac{1}{t} \int_0^t [a(\tau) + q(\tau)]\, d\tau.$$

Note that, actually, in the study of the variation of the characteristic exponents the inequality $\|Q(t)\| < \delta$ is essential not on the whole half-axis $\mathbb{R}_+$, but starting with some sufficiently large moment $T$. Indeed, the exponents are determined by the behavior of the solutions as $t \to \infty$, and the variation of the coefficients on a finite interval does not affect this behavior. Moreover, stable exponents do not change under perturbations such that $\|Q(t)\| \to 0$ as $t \to \infty$. Let us prove these statements.

THEOREM 5.2.1. *Let the characteristic exponents of system* (5.0.1) *be stable and*

$$\|Q(t)\| \to 0 \qquad as \qquad t \to \infty;$$

*then the spectra of systems* (5.0.1) *and* (5.0.4) *coincide.*

PROOF. Let

$$X(t) = \{x_1(t), \dots, x_n(t)\}$$

be a normal basis of system (5.0.1) and

$$\chi[x_i] = \lambda_i, \qquad i = 1, \dots, n.$$

Consider the system

$$\dot{x} = [A(t) + G_T(t)]x, \tag{5.2.2}$$

where the matrix $G_T(t) \equiv 0$ for $t \geqslant T$, i.e., the perturbation is restricted to the finite interval $[0, T]$. Take the basis $\widetilde{X}(t)$ of system (5.2.2) satisfying the initial condition $\widetilde{X}(T) = X(T)$. By virtue of the condition for $G_T(t)$, these bases coincide for $t \geqslant T$, i.e.,

$$\widetilde{X}(t) \equiv X(t), \qquad t \geqslant T. \tag{5.2.3}$$

The basis $\widetilde{X}(t)$ is normal since it is incompressible (because $X(t)$ is incompressible), consequently, it realizes the complete spectrum of system (5.2.2), and, by the condition (5.2.3), this spectrum coincides with that of system (5.0.1). Thus, a perturbation of a system on a finite interval of time does not change the spectrum.

We turn to the proof of the statement of the theorem. Let, to the contrary, the spectra of systems (5.0.3) and (5.0.4) be different for

$$\|Q(t)\| \to 0, \qquad t \to \infty,$$

i.e., let for some $j$

$$|\lambda_j - \lambda_j'| = a > 0. \tag{5.2.4}$$

Set $\varepsilon = a/2$ and, by the stability of characteristic exponents of system (5.0.1), define $\delta > 0$ such that for $\sup_{t\in\mathbb{R}_+} \|R(t)\| < \delta$ the spectrum

$$\alpha_1 \leqslant \alpha_2 \leqslant \cdots \leqslant \alpha_n$$

of the system

$$\dot{y} = [A(t) + R(t)]y$$

satisfy the condition (5.2.1),

$$|\alpha_i - \lambda_i| < a/2, \qquad i = 1, \dots, n. \tag{5.2.5}$$

For the same $\delta > 0$ we define $T > 0$ such that

$$\|Q(t)\| < \delta \qquad \text{for} \qquad t \geqslant T.$$

Let us construct the matrix $R(t)$ in the following way:

$$R(t) \equiv Q(t) \qquad \text{for} \qquad t \geqslant T,$$

and $R(t)$ is arbitrary with

$$\|R(t)\| < \delta \qquad \text{for} \qquad t \in [0, T].$$

According to the arguments at the beginning of the proof,

$$\alpha_i = \lambda_i', \qquad i = 1, 2, \ldots, n,$$

since the matrices $R(t)$ and $Q(t)$ differ on a finite interval of time; then the inequality (5.2.5) contradicts the inequality (5.2.4). □

REMARK 5.2.1. Without the assumption on the stability of characteristic exponents, Theorem 5.2.1 does not hold (see Perron's Example 4.4.1 and Example 5.1.6).

LEMMA 5.2.1. *The stability of characteristic exponents is invariant under Lyapunov transformations.*

PROOF. These transformations preserve both the spectrum of a system and the smallness of perturbations. □

Now we present a sufficient condition for the stability of characteristic exponents obtained by Malkin [**26**].

THEOREM 5.2.2 (Malkin). *Let the Cauchy matrix*

$$X(t, \tau) = \{x_1(t, \tau), \ldots, x_n(t, \tau)\}$$

*of system* (5.0.1) *be such that*

1. $\chi[x_i(t, \tau)] = \mu_i$, $i = 1, \ldots, n$,
2. *for any* $\gamma > 0$ *the following inequalities hold*:

$$\|x_i(t, \tau)\| \leqslant C \exp(\mu_i + \gamma)(t - \tau) \qquad \textit{for} \qquad t \geqslant \tau \geqslant 0, \tag{5.2.6}$$

$$\|x_i(t, \tau)\| \leqslant C \exp(\mu_i - \gamma)(t - \tau) \qquad \textit{for} \qquad \tau \geqslant t \geqslant 0, \tag{5.2.7}$$

*where $C$ is a constant depending only on $\gamma$ and independent of $\tau$ and $t$.*

*Then the characteristic exponents of system* (5.0.1) *are stable.*

REMARK 5.2.2. The notation $\mu_i$ for the exponents is introduced because we cannot guarantee that the Cauchy matrix is normal, and it may not realize the complete spectrum. Naturally, $\mu_i$ coincides with one of the $\lambda_j$, $i, j = 1, \ldots, n$.

PROOF. The proof of the theorem consists of three parts. Let us prove successively the following:

1. the shift of the characteristic exponents to the right is small,
2. system (5.0.1) is regular,
3. the shift of the characteristic exponents to the left is small.

We note that Malkin proved items 1 and 3 under the assumption that the system is regular. Bogdanov showed [7] that the conditions of the theorem guarantee that the system is regular; this result is presented in item 2.

1. Let $t_0 \in \mathbb{R}_+$. Any solution $y(t)$ of system (5.0.4), by virtue of the method of variation of constants, satisfies the integral equation

$$y(t) = X(t, t_0)y(t_0) + \int_{t_0}^{t} X(t, \tau)Q(\tau)y(\tau)\,d\tau.$$

Note that the equality $t_0 = \infty$ is possible if the integral converges. We distinquish $n$ solutions $y_1(t), \dots, y_n(t)$ of system (5.0.4), whose initial data at $t = t_0$ coincide with the initial data of the solutions $x_1(t), \dots, x_n(t)$ of system (5.0.1), constituting a normal basis of system (5.0.1). We assume that

$$\chi[x_k] = \lambda_k \leqslant \lambda_{k+1} = \chi[x_{k+1}], \qquad k = 1, \dots, n-1.$$

This gives $n$ integral equations of the form

$$y_k(t) = x_k(t) + \int_{t_0}^{t} X(t, \tau)Q(\tau)y_k(\tau)\,d\tau, \qquad k = 1, \dots, n. \tag{5.2.8}$$

Let us estimate the characteristic exponents of the solutions $y_k(t)$. Take an $\varepsilon > 0$ such that

$$0 < \varepsilon < (\lambda_n - \lambda_k)/2, \qquad \lambda_n \neq \lambda_k. \tag{5.2.9}$$

For such an $\varepsilon > 0$ there exists a constant $A$ such that

$$\|x_k(t)\| \leqslant Ae^{(\lambda_k+\varepsilon)t}, \qquad t \geqslant 0. \tag{5.2.10}$$

Our goal is to show that

$$\|y_k(t)\| \leqslant 2Ae^{(\lambda_k+\varepsilon)t}, \qquad t \geqslant 0. \tag{5.2.11}$$

Consider the Picard approximations for equation (5.2.8):

$$\begin{aligned} &y_k^{(0)}(t) = x_k(t), \\ &y_k^{(m)}(t) = x_k(t) + \int_{t_0}^{t} X(t, \tau)Q(\tau)y_k^{(m-1)}(\tau)\,d\tau, \qquad m = 1, 2, \dots. \end{aligned} \tag{5.2.12}$$

For $y_k^{(0)}$, the inequality (5.2.11) is satisfied by (5.2.10). Assume that for the $(m-1)$th approximation we have

$$\|y_k^{(m-1)}(t)\| < 2Ae^{(\lambda_k+\varepsilon)t}, \qquad t \geqslant 0, \tag{5.2.13}$$

and show that this estimate is valid for the $m$th approximation. From the formula (5.2.12) it follows that

$$\|y_k^{(m)}(t)\| \leqslant Ae^{(\lambda_k+\varepsilon)t} + \left| \int_{t_0}^{t} \|X(t, \tau)\| \cdot \|Q(\tau)\| \cdot \|y_k^{(m-1)}(\tau)\|\,d\tau \right|.$$

Let us estimate the integral, setting

$$t_0 = \begin{cases} 0, & \text{if } \lambda_n = \lambda_k, \\ \infty, & \text{if } \lambda_n > \lambda_k. \end{cases}$$

We assume that $\gamma = \varepsilon/2$ in the inequalities (5.2.6) and (5.2.7). Consider two cases:

a) $t_0 = 0$, i.e., $t \geqslant \tau$; using (5.2.6), we have

$$\int_0^t \|X(t,\tau)\| \cdot \|Q(\tau)\| \cdot \|y_k^{(m-1)}(\tau)\| \, d\tau \leqslant 2Ac\delta \int_0^t e^{(\lambda_n+\varepsilon/2)(t-\tau)} e^{(\lambda_k+\varepsilon)\tau} \, d\tau$$

$$= 2Ac\delta e^{(\lambda_n+\varepsilon/2)t} \int_0^t e^{\varepsilon\tau/2} \, d\tau < \frac{4Ac\delta}{\varepsilon} e^{(\lambda_n+\varepsilon)t},$$

b) $t_0 = \infty$, i.e., $\tau \geqslant t$; using (5.2.7), we obtain

$$\left| \int_\infty^t \|X(t,\tau)\| \cdot \|Q(\tau)\| \cdot \|y_k^{(m-1)}(\tau)\| \, d\tau \right| \leqslant 2Ac\delta \int_t^\infty e^{(\lambda_n-\varepsilon/2)(t-\tau)} e^{(\lambda_k+\varepsilon)\tau} \, d\tau$$

$$= 2Ac\delta e^{(\lambda_n-\varepsilon/2)t} \int_t^\infty e^{(\lambda_k-\lambda_n+3\varepsilon/2)\tau} \, d\tau$$

$$= \frac{2Ac\delta e^{(\lambda_n-\varepsilon/2)t} e^{(\lambda_k-\lambda_n+3\varepsilon/2)t}}{\lambda_n - \lambda_k - 3\varepsilon/2}$$

$$< \frac{4Ac\delta}{\varepsilon} e^{(\lambda_k+\varepsilon)t}.$$

The last two inequalities follow from the condition (5.2.9).

In both cases we have obtained identical estimates for the integral. The estimate (5.2.13) also holds for the $m$th step if the integral does not exceed the quantity $A\exp(\lambda_k + \varepsilon)t$. For this reason we choose $\delta < \varepsilon/(4C)$. Recall that $C$ depends on $\gamma = \varepsilon/2$.

We know **[5]** that the Picard process converges to the solution of the integral equation (5.2.8) for all $t \in \mathbb{R}_+$; therefore, from (5.2.13), as $m \to \infty$, we obtain that the estimate is valid.

Thus, we have $n$ solutions $y_1(t), \dots, y_n(t)$ of system (5.0.4). Let us show that they form a basis of this system. Indeed, as can be seen from the estimates for the integrals, for $t = 0$ and sufficiently small $\delta$ the vectors $y_1(0), y_2(0), \dots, y_n(0)$ differ little from the vectors $x_1(0), x_2(0), \dots, x_n(0)$, respectively . But the latter are linearly independent; therefore,

$$\operatorname{Det}\{x_1(0), \dots, x_n(0)\} \neq 0;$$

this is sufficient for the solutions $y_1(t), \dots, y_n(t)$ to be linearly independent. Moreover, from the estimates (5.2.11) we have

$$\chi[y_k] \leqslant \lambda_k + \varepsilon.$$

If the basis $y_1(t), \dots, y_n(t)$ is not normal, then, by passing to a normal basis, the exponents can only diminish; therefore, we have

$$\lambda'_k \leqslant \lambda_k + \varepsilon, \qquad k = 1, 2, \dots, n. \tag{5.2.14}$$

2. Let us show that under our assumptions system (5.0.1) satisfies the conditions of Lemma 3.5.1 on regularity; i.e.,

1) there exists $\lim_{t\to\infty} \frac{1}{t} \int_0^t \operatorname{Re} \operatorname{Sp} A(\tau) \, d\tau = S$,

2) $\sum_{i=1}^n \mu_i = S$.

According to the Ostrogradskiĭ-Liouville formula, we have

$$\operatorname{Det} X(t,0) = \exp \int_0^t \operatorname{Sp} A(\tau) \, d\tau.$$

Note that

$$[\operatorname{Det} X(t,0)]^{-1} = \operatorname{Det} X^{-1}(t,0) = \operatorname{Det} X(0,t).$$

This and the condition (5.2.7) imply that

$$e^{-\int_0^t \operatorname{Sp} A(\tau)\,d\tau} = \operatorname{Det} X(0,t) \leqslant C^n n! \exp\left(\sum_{i=1}^n (\mu_i - \gamma)(0-t)\right)$$
$$= C^n n! \exp\left(-\sum_{i=1}^n \mu_i t\right)\exp(n\gamma t).$$

The right-hand side of the inequality has a sharp characteristic exponent; therefore,

$$\overline{\lim_{t\to\infty}} \frac{1}{t}\left(-\int_0^t \operatorname{Re} \operatorname{Sp} A(\tau)\,d\tau\right) \leqslant -\sum_{i=1}^n \mu_i + n\gamma,$$

or

$$\underline{\lim_{t\to\infty}} \frac{1}{t}\int_0^t \operatorname{Re} \operatorname{Sp} A(\tau)\,d\tau \geqslant \sum_{i=1}^n \mu_i - n\gamma. \tag{5.2.15}$$

By the Lyapunov inequality (2.5.1) and by virtue of (5.2.15), we have

$$\begin{aligned}\sum_{i=1}^n \mu_i &\geqslant \overline{\lim_{t\to\infty}} \frac{1}{t}\int_0^t \operatorname{Re} \operatorname{Sp} A(\tau)\,d\tau \\ &\geqslant \underline{\lim_{y\to\infty}} \frac{1}{t}\int_0^t \operatorname{Re} \operatorname{Sp} A(\tau)\,d\tau \\ &\geqslant \sum_{i=1}^n \mu_i - n\gamma.\end{aligned} \tag{5.2.16}$$

Since $\gamma$ is arbitrarily small, the inequalities (5.2.16) become a chain of equalities and both statements formulated at the beginning of the proof of this item are valid.

3. Now let us show that the shift of the exponents to the left is small. The regularity of system (5.0.1) implies that

$$\sum_{i=1}^n \lambda_i = \chi\left[\exp\int_0^t \operatorname{Sp} A(\tau)\,d\tau\right].$$

By the Lyapunov inequality, we have

$$\sum_{i=1}^n \lambda_i' \geqslant \chi\left[\exp\int_0^t (\operatorname{Sp} A(\tau) + \operatorname{Sp} Q(\tau))\,d\tau\right] \tag{5.2.17}$$

for system (5.0.4).

Under the condition (5.0.5),

$$\left|\chi \exp\int_0^T \operatorname{Sp} Q(\tau)\,d\tau\right| \leqslant n\delta.$$

The function

$$\exp\left[\int_0^t \operatorname{Sp} A(\tau)\,d\tau\right]$$

has a sharp charactertistic exponent; therefore, (5.2.17) implies

$$\sum_{i=1}^{n} \lambda_i' \geqslant \chi\left[\exp \int_0^t \operatorname{Sp} A(\tau)\, d\tau\right] + \chi\left[\exp \int_0^t \operatorname{Sp} Q(\tau)\, d\tau\right]$$

(see Theorem 2.1.1), and, finally,

$$\sum_{i=1}^{n} \lambda_i' \geqslant \sum_{i=1}^{n} \lambda_i - n\delta. \tag{5.2.18}$$

According to $\varepsilon > 0$ we choose a $\delta > 0$ such that (5.2.14) is satisfied and introduce $\gamma_i > 0$ so that

$$\lambda_i' = \lambda_i + \varepsilon - \gamma_i, \qquad i = 1, \dots, n. \tag{5.2.19}$$

Substitute the expression obtained for $\lambda_i'$ in (5.2.18):

$$\sum_{i=1}^{n} \lambda_i + n\varepsilon - \sum_{i=1}^{n} \gamma_i \geqslant \sum_{i=1}^{n} \lambda_i - n\delta, \qquad \text{or} \qquad \gamma_i \leqslant n(\varepsilon + \delta).$$

From the last condition and (5.2.19), we have

$$\lambda_i' \geqslant \lambda_i + \varepsilon - n(\varepsilon + \delta). \tag{5.2.20}$$

Combining (5.2.14) and (5.2.20), we obtain our claim. □

COROLLARY 5.2.1. *The Cauchy matrix $X(t, \tau)$ satisfying the conditions* (5.2.6) *and* (5.2.7) *is a normal fundamental matrix.*

PROOF. For this matrix the inequalities (5.2.16) imply that the Lyapunov inequality (2.5.1) is satisfied; consequently, the matrix is normal. □

THEOREM 5.2.3. *Linear systems with constant coefficients have stable characteristic exponents.*

PROOF. Let system (5.0.1) have the matrix of coefficients $A \equiv \text{const}$ (such systems were considered in Chapter I in more detail). The Cauchy matrix for such a system has the form

$$X(t, \tau) = \underset{\tau}{\overset{t}{\Omega}} A = e^{A(t-\tau)}.$$

Let us introduce $S$, i.e., the matrix that transforms $A$ to the Jordan form,

$$B = S^{-1} A S = \operatorname{diag}[J_{p_1}(\lambda_1), \dots, J_{p_k}(\lambda_k)],$$

where $\lambda_1, \dots, \lambda_k$ are the eigenvalues of the matrix $A$ (not necessarily distinct); $J_\nu(\lambda)$ is the Jordan block of order $\nu$, $\sum_{i=1}^{k} p_i = n$. Hence,

$$e^{A(t-\tau)} = S e^{B(t-\tau)} S^{-1}.$$

Moreover,

$$e^{B(t-\tau)} = \operatorname{diag}\left[e^{J_{p_1}(\lambda_1)(t-\tau)}, \dots, e^{J_{p_k}(\lambda_k)(t-\tau)}\right], \tag{5.2.21}$$

where

$$e^{J_\nu(\lambda)(t-\tau)} = e^{\lambda(t-\tau)} \begin{pmatrix} 1 & \dots & & 0 \\ t-\tau & & & \vdots \\ \vdots & \ddots & & \vdots \\ \dfrac{(t-\tau)^{\nu-1}}{(\nu-1)!} & \dots & t-\tau & 1 \end{pmatrix}.$$

Each element of the matrix $\exp B(t-\tau)$ satisfies the estimates (5.2.6) and (5.2.7). Indeed, it has the following general form:

$$\frac{1}{k!}(t-\tau)^k e^{\lambda(t-\tau)}, \qquad k \in \{0, 1, \dots, n-1\}. \tag{5.2.22}$$

The inequalities (5.2.6) and (5.2.7) for this function have the form

$$\begin{aligned} \left|\frac{1}{k!}(t-\tau)^k e^{\lambda(t-\tau)}\right| &\leqslant Ce^{(\operatorname{Re}\lambda+\gamma)(t-\tau)}, \qquad t \geqslant \tau \geqslant 0, \\ \left|\frac{1}{k!}(t-\tau)^k e^{\lambda(t-\tau)}\right| &\leqslant Ce^{(\operatorname{Re}\lambda-\gamma)(t-\tau)}, \qquad \tau \geqslant t \geqslant 0. \end{aligned} \tag{5.2.23}$$

Dividing by the exponential, we obtain

$$\begin{aligned} \left|\frac{1}{k!}(t-\tau)^k\right| &\leqslant Ce^{\gamma(t-\tau)}, \qquad t \geqslant \tau \geqslant 0, \\ \left|\frac{1}{k!}(t-\tau)^k\right| &\leqslant Ce^{\gamma(\tau-t)}, \qquad \tau \geqslant t \geqslant 0. \end{aligned}$$

Both inequalities are reduced to a single one, and, obviously, there exists a constant $C$ depending on $\gamma$ and independent of $\tau$ that realizes the inequality $\frac{1}{k!}\theta^k \leqslant C\exp\gamma\theta$, $\theta \geqslant 0$, namely

$$C = \max_{\theta\in\mathbb{R}_+} \frac{1}{k!}\theta^k \exp(-\gamma\theta).$$

We return to the matrix (5.2.21):

$$e^{A(t-\tau)} = V(t)S^{-1} = \{v_1^{(1)}, \dots, v_{p_1}^{(1)}, v_1^{(2)}, \dots, v_{p_2}^{(2)}, \dots, v_1^{(k)}, \dots, v_{p_k}^{(k)}\}S^{-1}.$$

The columns of the matrix $V(t)$ are solutions of system (5.0.1). They are divided into $k$ groups. The first one is the result of multiplication of the matrix $S$ by the first $p_1$ columns of the matrix $\exp B(t-\tau)$, etc., and the last one is the result of multiplication of $S$ by the last $p_k$ columns of the matrix $\exp B(t-\tau)$. The solutions of the $m$th group have the characteristic exponent

$$\operatorname{Re}\lambda_m, \qquad m = 1, 2, \dots, k.$$

Multiplying the matrix $V(t)$ by $S^{-1}$ from the right, we have

$$e^{A(t-\tau)} = \{x_1(t,\tau), x_2(t,\tau), \dots, x_n(t,\tau)\},$$

where each vector $x_i(t,\tau)$ is a linear combination of the solutions $v_1(t), \dots, v_n(t)$. Therefore, the components of any solution $x_i(t,\tau)$ represent linear combinations of functions of the form (5.2.22) with coefficients depending on the constant matrices $S$ and $S^{-1}$. For each of these functions, the estimates (5.2.23) are satisfied and the inequalities can only be strengthened if in the right-hand side $\operatorname{Re}\lambda$ is replaced with the maximal exponent of the linear combination, i.e., with the exponent of the solution

$x_i(t,\tau)$. The constant $C$ changes its value because of the multiplication by the constant matrices $S$ and $S^{-1}$. Estimating the vectors $x_i(t,\tau)$, $i = 1,\dots,n$, componentwise, we verify that the inequalities (5.2.6) and (5.2.7) hold. □

THEOREM 5.2.4. *The characteristic exponents of systems that are reducible to autonomous ones are stable.*

PROOF. According to Erugin's Theorem 3.2.1, a reducible system has a fundamental matrix of the form

$$X(t) = L(t)\exp At,$$

where $L(t)$ is Lyapunov, and $A$ is constant. Thus,

$$X(t,\tau) = L(t)e^{A(t-\tau)}L^{-1}(\tau).$$

It remains to repeat the proof of the previous theorem, changing $S$ to $L(t)S$, and $S^{-1}$ to $S^{-1}L^{-1}(t)$. Although these matrices are not constant, they are Lyapunov, i.e., bounded on $\mathbb{R}_+$; this affects the value of the constant $C$, but not the form of the estimates (5.2.6) and (5.2.7). □

THEOREM 5.2.5. *The characteristic exponents of systems that are almost reducible to autonomous ones are stable.*

PROOF. Let system (5.0.1) be almost reducible to the system

$$\dot{y} = By, \qquad B = \text{const}, \tag{5.2.24}$$

with the characteristic exponents

$$\mu_1 \leqslant \mu_2 \leqslant \cdots \leqslant \mu_n.$$

This means that for any $\alpha > 0$ there exists a Lyapunov matrix $L_\alpha(t)$ such that the system

$$\dot{z} = (L_\alpha^{-1}AL_\alpha - L_\alpha^{-1}\dot{L}_\alpha)z = [B + \Phi(t)]z \tag{5.2.25}$$

has the spectrum

$$\lambda_1 \leqslant \lambda_2 \leqslant \cdots \leqslant \lambda_n,$$

and

$$\sup_{t\geqslant 0} \|\Phi(t)\| \leqslant \alpha.$$

By the stability of the exponents of system (5.2.24), for any $\varepsilon > 0$ there exists a $\delta > 0$ such that the characteristic exponents $\nu_1 \leqslant \nu_2 \leqslant \cdots \leqslant \nu_n$ of the system

$$\dot{V} = [B + \widetilde{Q}(t)]V, \qquad \sup_{t\geqslant 0} \|\widetilde{Q}(t)\| < \Delta, \tag{5.2.26}$$

satisfy the inequalities

$$|\nu_i - \mu_i| < \varepsilon/2, \qquad i = 1,\dots,n.$$

Take $\alpha = \Delta/2$; applying the transformation $y = L_\alpha(t)w$ to system (5.0.4), we obtain the system

$$\dot{w} = (B + \Phi(t) + L_\alpha^{-1}QL_\alpha)w \tag{5.2.27}$$

with the spectrum

$$\lambda_1' \leqslant \lambda_2' \leqslant \cdots \leqslant \lambda_n'.$$

Let
$$\|L_\alpha(t)\| \leqslant K, \qquad \|L_\alpha^{-1}(t)\| \leqslant K \qquad \text{for} \qquad t \in \mathbb{R}_+;$$
such a $K$ exists, since $L_\alpha(t)$ is Lyapunov; we choose $\delta = \Delta/(2K^2)$ in (5.0.5). Then for system (5.2.27) we have

$$\|\Phi(t) + L_\alpha^{-1}Q(t)L_\alpha\| \leqslant \|\Phi(t)\| + \|L_\alpha^{-1}QL_\alpha\| < \frac{\Delta}{2} + K^2\delta = \frac{\Delta}{2} + \frac{\Delta}{2} = \Delta.$$

Both systems (5.2.27) and (5.2.25) are particular cases of system (5.2.26); thus,

$$|\lambda_i - \mu_i| < \varepsilon/2, \qquad |\lambda_i' - \mu_i| < \varepsilon/2, \qquad i = 1, \dots, n;$$

consequently,
$$|\lambda_i - \lambda_i'| < \varepsilon.$$
Hence, for $\varepsilon > 0$ we have found a corresponding $\Delta > 0$, and for it we have defined $\alpha = \Delta/2$ and the matrix $L_\alpha(t)$, generating the constant $K$, and, finally, $\delta = \Delta/(2K^2)$.□

THEOREM 5.2.6. *Let $A_1(t)$ and $A_2(t)$ be bounded and continuous matrices on $\mathbb{R}_+$ such that*

1.

(5.2.28)
$$\|A_1(t) - A_2(t)\| \underset{t\to\infty}{\longrightarrow} 0,$$

2. *one of the systems*

$$\dot{x} = A_1(t)x,$$
$$\dot{x} = A_2(t)x$$

*is autonomous, or reducible to an autonomous one, or is almost reducible to an autonomous one.*

*Then the characteristic exponents of these systems are stable and coincide.*

PROOF. Let the system $\dot{x} = A_1(t)x$ possess one of the properties indicated in item 2. We write the second system in the form

$$\dot{x} = [A_1(t) + (A_2(t) - A_1(t))]x.$$

The final statement of the theorem follows from the stability of the exponents of the first system (Theorems 5.2.3, 5.2.4, 5.2.5) and Theorem 5.2.1, whose conditions are satisfied by virtue of (5.2.28). □

EXAMPLE 5.2.2. Consider the system

$$\dot{x}_1 = (1 + \cos t)x_1 + \frac{1}{(t+1)^2}x_2,$$
$$\dot{x}_2 = e^{-2t}x_1 + x_2 \sin t.$$

The characteristic exponents of this system are stable and equal to $0, 1$. Indeed, compare our system with the system

$$\dot{x}_1 = (1 + \cos t)x_1,$$
$$\dot{x}_2 = x_2 \sin t,$$

which is $2\pi$-periodic and, consequently, reducible to a system with constant coefficients. Its exponents

$$\lambda_1 = \lim_{t\to\infty} \frac{1}{t} \int_0^t (1+\cos\tau)\,d\tau = 1,$$

$$\lambda_2 = \lim_{t\to\infty} \frac{1}{t} \int_0^t \sin\tau\,d\tau = 0$$

are stable. Here we have used Theorems 5.2.4 and 5.2.6.

## §3. Integral separateness

Let us touch upon the history of the problem **[23]**. Perron **[32]** obtained a result which forms the basis of the theory of stability of exponents. Let us describe it in our terms. Let system (5.0.1) be diagonal, i.e., let

$$\dot{x} = \operatorname{diag}[a_1(t), \dots, a_n(t)]x,$$

where

$$\operatorname{Re}(a_{k+1}(t) - a_k(t)) \geqslant a > 0, \qquad k = 1, \dots, n-1, \quad t \geqslant 0; \tag{5.3.1}$$

moreover, let $\|Q(t)\| \to 0$ as $t \to \infty$. Then the characteristic exponents of systems (5.0.1) and (5.0.4) coincide, i.e.,

$$\lambda_i' = \lambda_i = \overline{\lim_{t\to\infty}} \frac{1}{t} \int_0^t \operatorname{Re} a_i(\tau)\,d\tau, \qquad i = 1, \dots, n.$$

The inequality (5.3.1) is called the *condition for separateness of the diagonal* of system (5.0.1), or the *condition for separateness of the functions* $a_1(t), \dots, a_n(t)$. Note that the Perron conditions allow one to calculate the characteristic exponents via the coefficients.

EXAMPLE 5.3.1. Let us find the characteristic exponents of the system

$$\dot{x}_1 = \left(\cos\ln t + \sin\ln t + \frac{2}{t+1}\right) x_1 + te^{-t}x_2,$$

$$\dot{x}_2 = \frac{\sin\ln t}{(t-1)^2}x_1 + 2x_2.$$

The matrix of coefficients of the system is the sum of the matrices

$$\operatorname{diag}[\cos\ln t + \sin\ln t, 2] \qquad \text{and} \qquad Q(t) = \begin{pmatrix} \dfrac{2}{t+1} & te^{-t} \\ \dfrac{\sin\ln t}{(t-1)^2} & 0 \end{pmatrix};$$

here $\|Q(t)\| \to 0$ as $t \to \infty$. Note that

$$2 - (\sin\ln t + \cos\ln t) \geqslant 2 - \sqrt{2} > 0,$$

i.e., the condition for separateness is satisfied; therefore,

$$\lambda_1 = \overline{\lim_{t\to\infty}} \frac{1}{t} \int_0^t (\sin\ln\tau + \cos\ln\tau)\,d\tau = 1, \qquad \lambda_2 = 2.$$

Let us see whether it is possible to obtain this result by the methods considered in the previous section. If it turns out that the exponents of the system

$$\dot{x} = \operatorname{diag}[(\sin \ln t + \cos \ln t), 2]x$$

are stable, then it is possible. It is clear that the diagonal system is neither autonomous, nor reducible, nor almost reducible to an autonomous system. Let us verify that the condition (5.2.6) of Theorem 5.2.2 is satisfied for $x_1(t)$; we have

$$|x_1(t, \tau)| = \exp(t \sin \ln t - \tau \sin \ln \tau).$$

We must show that for any $\gamma > 0$ there exists a constant $C(\gamma)$ such that for $t \geqslant \tau \geqslant 0$ the inequality

$$e^{t \sin \ln t - \tau \sin \ln \tau} \leqslant Ce^{(1+\gamma)(t-\tau)}$$

holds, or

$$t \sin \ln t - \tau \sin \ln \tau \leqslant (1+\gamma)(t-\tau) + \ln C.$$

Take $\gamma = e^{-\pi}(1-e^{-\pi})^{-1}$ and consider the sequences

$$t_m = \exp\left(2m\pi + \frac{\pi}{2}\right), \qquad \tau_m = \exp\left(2m\pi - \frac{\pi}{2}\right).$$

Hence,

$$e^{2m\pi+\frac{\pi}{2}} + e^{2m\pi-\frac{\pi}{2}} \leqslant (1+\gamma)(e^{2m\pi+\frac{\pi}{2}} - e^{2m\pi-\frac{\pi}{2}}) + \ln C,$$

or

$$1 + e^{-\pi} \leqslant \left(1 + \frac{e^{-\pi}}{1-e^{-\pi}}\right)(1-e^{-\pi}) + e^{-2m\pi} \ln C;$$

therefore,

$$1 + e^{-\pi} \leqslant 1 + e^{-2m\pi} \ln C.$$

Obviously, there is no constant $C$ such that the inequality holds for arbitrarily large $m$, i.e., the sufficient condition for stability is not satisfied.

In what follows, we shall see that, under the condition (5.3.1), the exponents of a diagonal system are actually stable, and, at the same time, the condition for separateness of the diagonal can be substituted by the condition for integral separateness.

DEFINITION 5.3.1. The bounded continuous functions $a_1(t), a_2(t), \ldots, a_n(t)$ are said to be *integrally separated* on $\mathbb{R}_+$ if there exist constants $a > 0$ and $d \geqslant 0$ such that

$$\int_s^t [a_{i+1}(u) - a_i(u)]\, du \geqslant a(t-s) - d \tag{5.3.2}$$

for all $t \geqslant s \geqslant 0$, $i = 1, \ldots, n-1$.

Obviously, the condition for separateness implies integral separateness, but not vice versa.

EXAMPLE 5.3.2. The functions $\cos t, 1$ are not separated, but are integrally separated with the constants $a_1 = 1$, $d = 2$. It should be noted that Perron's result is not applicable to the diagonal system

$$\dot{x} = \operatorname{diag}[\cos t, 1]x,$$

while Theorem 5.2.6 is, in contrast, applicable. Indeed, if the coefficients of this system are perturbed by small summands tending to zero as $t \to \infty$, then, according to Theorem 5.2.6, the characteristic exponents of both systems are stable and coincide, i.e., are equal to 0 and 1. This statement follows from the fact that the unperturbed system is reducible to an autonomous one.

The notion of integral separateness was introduced for systems of a general form by Bylov [**11**]. It, naturally, appeals to properties of fundamental matrices and not to the coefficients.

DEFINITION 5.3.2. A linear system is said to be a *system with integral separateness* if it has solutions $x_1(t), \dots, x_n(t)$ such that the inequality

$$\frac{\|x_{i+1}(t)\|}{\|x_{i+1}(s)\|} : \frac{\|x_i(t)\|}{\|x_i(s)\|} \geqslant de^{a(t-s)}, \qquad i = 1, \dots, n-1, \tag{5.3.3}$$

with some constants $a > 0$, $d \geqslant 1$, is valid for all $t \geqslant s$.

Integral separateness is rather a strict condition and, therefore, has a lot of implications concerning different properties of a system, in particular, as will be shown below, the stability of characteristic exponents. Note some obvious ones.

PROPERTY 5.3.1. *Integrally separated systems have different characteristic exponents.*

PROOF. We set $s = 0$ in the inequality (5.3.3) and take logarithms:

$$\ln \|x_{i+1}(t)\| - \ln \|x_i(t)\| \geqslant at + \ln d - \ln \|x_i(0)\| + \ln \|x_{i+1}(0)\|,$$

whence

$$\varliminf_{t\to\infty} \frac{1}{t}(\ln \|x_{i+1}(t)\| - \ln \|x_i(t)\|) \geqslant a.$$

At the same time,

$$\begin{aligned} \varliminf_{t\to\infty} \frac{1}{t} \ln \frac{\|x_{i+1}(t)\|}{\|x_i(t)\|} &\leqslant \varlimsup_{t\to\infty} \frac{1}{t} \ln \|x_{i+1}(t)\| + \varliminf_{t\to\infty} \frac{1}{t} \ln \frac{1}{\|x_i(t)\|} \\ &= \chi[x_{i+1}] - \chi[x_i]. \end{aligned}$$

Finally, we have

$$\chi[x_{i+1}] - \chi[x_i] \geqslant a > 0.$$

Note that the fact that characteristic exponents are different does not necessarily imply integral separateness of solutions.

EXAMPLE 5.3.3. The system

$$\begin{aligned} \dot{x}_1 &= 0, \\ \dot{x}_2 &= (\sin \ln t + \cos \ln t)x_2 \end{aligned}$$

has the fundamental matrix

$$X(t) = \{x_1(t), x_2(t)\} = \begin{pmatrix} 1 & 0 \\ 0 & e^{t \sin \ln t} \end{pmatrix}.$$

Therefore,

$$\chi[x_1] = 0, \qquad \chi[x_2] = 1.$$

Let us verify whether the condition (5.3.3) is satisfied:

$$\frac{\|x_2(t)\|}{\|x_2(s)\|} : \frac{\|x_1(t)\|}{\|x_1(s)\|} = e^{t \sin \ln t - s \sin \ln s}.$$

Take

$$t = \exp 2m\pi, \qquad s = \exp(2m - l)\pi, \qquad 2m > l \geqslant 1.$$

For these values of $t$ and $s$ the right-hand side is always equal to 1; therefore, (5.3.3) is not satisfied.

PROPERTY 5.3.2. *Integral separateness is invariant under Lyapunov transformations.*

Let an integrally separated system (5.0.1) be reduced to the system $\dot{y} = B(t)y$ by a Lyapunov transformation $y = L(t)x$. We show that this system is also integrally separated. Consider its basis

$$Y(t) = \{L(t)x_1(t), L(t)x_2(t), \dots, L(t)x_n(t)\},$$

where $x_1(t), \dots, x_n(t)$ are solutions of (5.0.1) satisfying (5.3.3). Let a constant $K \geqslant 1$ be such that

$$\|L(t)\| \leqslant K, \qquad \|L^{-1}(t)\| \leqslant K \qquad \text{for} \qquad t \in \mathbb{R}_+.$$

Consider the following inequalities:

$$\|L^{-1}Lx\| \leqslant \|L^{-1}\|\|Lx\| \leqslant K\|Lx\|;$$

therefore,

$$\|Lx\| \geqslant \frac{\|x\|}{K}, \qquad \|Lx\| \leqslant \|L\|\|x\|.$$

Now let us verify directly the inequality (5.3.3):

$$\frac{\|L(t)x_{i+1}(t)\| \cdot \|L(s)x_i(s)\|}{\|L(s)x_{i+1}(s)\| \cdot \|L(t)x_i(t)\|} \geqslant \frac{1}{K^4} \frac{\|x_{i+1}(t)\| \cdot \|x_i(s)\|}{\|x_{i+1}(s)\| \cdot \|x_i(t)\|} \geqslant \frac{d}{K^4} e^{a(t-s)}.$$

PROPERTY 5.3.3. *A basis of system* (5.0.1) *satisfying the inequality* (5.3.3) *is normal.*

Indeed, all the characteristic exponents of the solutions $x_1(t), \dots, x_n(t)$ are different, i.e., a linear combination whose exponent is less than those of the solutions is impossible.

PROPERTY 5.3.4. *The diagonal system*

$$\dot{x} = \operatorname{diag}[a_1(t), \dots, a_n(t)]x$$

*is integrally separated if its diagonal coefficients are integrally separated.*

To prove this statement, it is sufficient to consider the basis

$$X(t) = \operatorname{diag}\left[\exp \int_0^t a_1\, d\tau, \dots, \exp \int_0^t a_n\, d\tau\right]$$

and verify whether the inequalities (5.3.3) are satisfied for it if the inequalities (5.3.2) hold.

Note that the converse for the latter statement is also valid; this immediately follows from the subsequent results of this section.

The following important property was proved by Bylov **[11]**.

THEOREM 5.3.1 (Bylov). *An integrally separated system is reducible to a diagonal one.*

PROOF. Let the inequality (5.3.3) be satisfied for the solutions $x_1(t), \dots, x_n(t)$ of system (5.0.1). For the sake of convenience, we introduce exponential characteristics of solutions in the following way: we associate the function

$$p_x(t) = \frac{d}{dt} \ln \|x(t)\|$$

to the vector-function $x(t)$; this defines functions $p_1(t), \dots, p_n(t)$, and

$$\|x_i(t)\| = \|x_i(s)\| \exp \int_0^t p_i(\tau)\, d\tau, \qquad i = 1, \dots, n. \tag{5.3.4}$$

From the conditions (5.3.3) we have

$$\int_s^t (p_{i+1}(\tau) - p_i(\tau))\, d\tau \geqslant a(t-s) + \ln d, \tag{5.3.5}$$

i.e., we come to the integral separateness of the set of functions $p_1(t), \dots, p_n(t)$. Note that $p_x(t)$ is the same for $x(t)$ and $Cx(t)$, $C$ a constant.

Now let us turn to Theorem 3.3.3 on the reduction of a system to block-triangular form. According to Corollary 3.3.2, a linear system is reducible to a diagonal one by means of a Lyapunov transformation if and only if it has a basis

$$X(t) = \{x_1(t), \dots, x_n(t)\}$$

satisfying the condition

$$\frac{G(X)}{\|x_1(t)\|^2 \cdot \|x_2(t)\|^2 \cdots \|x_n(t)\|^2} \geqslant \rho > 0 \qquad \text{for} \qquad t \in \mathbb{R}_+,$$

where $G(X)$ is the Gram determinant of the solutions $x_1(t), \dots, x_n(t)$.

The same condition in Remark 3.3.4 is reformulated in terms of angles between subspaces,

$$\frac{G(X)}{\|x_1\|^2 \cdot \|x_2\|^2 \cdots \|x_n\|^2} = \sin^2 \beta_1 \cdots \sin^2 \beta_{n-1} \geqslant \rho > 0, \tag{5.3.6}$$

where

$$\beta_k = \sphericalangle(L_k, x_{k+1}), \qquad k = 1, \dots, n-1,$$

and $L_k$ is the $k$-dimensional linear subspace spanned by the solution vectors

$$x_1(t), x_2(t), \dots, x_k(t).$$

Thus, our goal is to prove that all the angles $\beta_k$ are bounded away from zero.

Moreover, we show that for all $x \in L_k$ the inequality

$$\frac{\|x(t)\|}{\|x(s)\|} \leqslant B_k e^{\int_s^t p_k(\tau)\, d\tau}, \qquad t \geqslant s \geqslant 0, \tag{5.3.7}$$

is valid. From what follows it is clear that this inequality provides the property required for the angles.

We introduce the assumption

$$\inf_{i, t \in \mathbb{R}_+} p_i(t) = \lambda > 0, \tag{5.3.8}$$

which will be dropped at the end of the proof.

We use induction. For $k = 1$, the statement of the theorem and the estimate (5.3.7) are obvious. Now assume that for all $k = 1, 2, \dots, m$ the inequality

$$\frac{G(x_1, \dots, x_k)}{\|x_1\|^2 \cdots \|x_k\|^2} = \sin^2 \beta_1 \cdots \sin^2 \beta_{k-1} \geqslant \rho > 0 \tag{5.3.9}$$

has been established, and for $x \in L_k$ the estimate

$$\frac{\|x(t)\|}{\|x(s)\|} \leqslant B_k \exp \int_s^t p_k(\tau)\, d\tau \tag{5.3.10}$$

holds.

We show that both statements are preserved for $k = m + 1$. Let, contrariwise, the first condition be violated. Then there exists a sequence $t_i \to \infty$ as $i \to \infty$ and a sequence of solutions $x_i'(t_i) \in L_m$ such that

$$\beta_m(t_i) = \sphericalangle\{x_i'(t_i), x_{m+1}(t_i)\} \underset{i \to \infty}{\longrightarrow} 0. \tag{5.3.11}$$

Without loss of generality, we assume that

$$\|x_i'(t_i)\| = \|x_{m+1}(t_i)\| = 1.$$

Therefore,

$$\|x_{m+1}(t_i) - x_i'(t_i)\| \to 0 \qquad \text{as} \qquad i \to \infty,$$

and, according to the theorem on integral continuity **[5]**, for any fixed $T > 0$ we have

$$\|x_{m+1}(t_i + T) - x_i'(t_i + T)\| \underset{i \to \infty}{\longrightarrow} 0. \tag{5.3.12}$$

At the same time, by virtue of the assumption (5.3.10) and the identity (5.3.4), we have

$$\begin{aligned}
&\|x_{m+1}(t_i + T) - x_i'(t_i + T)\| \\
&\quad \geqslant \|x_{m+1}(t_i + T)\| - \|x_i'(t_i + T)\| \\
&\quad \geqslant \exp\left(\int_{t_i}^{t_i+T} p_{m+1}(\tau)\, d\tau\right) - B_m \exp\left(\int_{t_i}^{t_i+T} p_m(\tau)\, d\tau\right) \\
&\quad = \exp\left(\int_{t_i}^{t_i+T} p_{m+1}(\tau)\, d\tau\right)\left[1 - B_m \exp\left(\int_{t_i}^{t_i+T} (p_m(\tau) - p_{m+1}(\tau))\, d\tau\right)\right] \\
&\quad \geqslant e^{\lambda T}\left[1 - \frac{B_m}{d} e^{-aT}\right] > 1.
\end{aligned}$$

The last inequality can always be ensured by the choice of a sufficiently large $T > 0$. The inequality obtained contradicts the condition (5.3.12), which is due to the assumption (5.3.11). Thus, the inequality (5.3.9) holds for $k = m + 1$.

We pass to the proof of the inequality (5.3.10) for $k = m + 1$. Let

$$x \in L_{m+1}, \qquad \|x\| \neq 0.$$

Represent this solution in the form

$$\frac{x(t)}{\|x(s)\|} = C_1 x'(t) + C_2 x_{m+1}(t), \tag{5.3.13}$$

where $x' \in L_m$ and

$$\|x'(s)\| = \|x_{m+1}(s)\| = 1,$$

$C_1$ and $C_2$ do not vanish simultaneously. By virtue of the above arguments,

$$0 < \rho \leqslant \sphericalangle(x'(s), x_{m+1}(s)) \leqslant \pi/2. \tag{5.3.14}$$

By the representation (5.3.13) for $t = s$, we have that a linear combination of two unit vectors forming an angle satisfying the condition (5.2.14) is a unit vector. Obviously, $|C_1| \leqslant C$, $|C_2| \leqslant C$ and the constant $C$ does not depend either on $x \in L_{m+1}$ or on $s \geqslant 0$. Therefore,

$$\begin{aligned}
\frac{\|x(t)\|}{\|x(s)\|} &\leqslant CB_m \exp \int_s^t p_m(\tau)\,d\tau + C \exp \int_s^t p_{m+1}(\tau)\,d\tau \\
&\leqslant \exp \int_s^t p_{m+1}(\tau)\,d\tau \left[C + CB_m \exp \int_s^t (p_m - p_{m+1})\,d\tau\right] \\
&\leqslant C\left[1 + \frac{B_m}{d} e^{-a(t-s)}\right] \exp \int_s^t p_{m+1}\,d\tau \\
&< C\left[1 + \frac{B_m}{d}\right] \exp \int_s^t p_{m+1}(\tau)\,d\tau.
\end{aligned}$$

Thus, we have completed the proof under the assumption (5.3.8).

The case $\inf_{i,t\in\mathbb{R}_+} p_i(t) \leqslant 0$ is reduced to the strict inequality (5.3.8) by the $\Lambda$-transformation $z = x \exp \Lambda t$, where $\Lambda$ is sufficiently large and satisfies

$$\Lambda + \inf_{i,t} p_i > 0.$$

This transformation changes the norm, increases the characteristic exponents by $\Lambda$, but does not affect in any way the angles; therefore, the condition (5.3.6) remains valid. □

COROLLARY 5.3.1. *If a system is integrally separated, then any solution $x \in L_k$, where $L_k$ is the linear supspace spanned by the solutions $x_1(t), \dots, x_k(t)$ from (5.3.3), satisfies the estimate*

$$\|x(t)\| \leqslant \|x(s)\| B_k e^{\int_s^t p_k(\tau)\,d\tau}, \tag{5.3.15}$$

*where $B_k$ is independent of $s$.*

COROLLARY 5.3.2. *An integrally separated system is reducible to the diagonal system*

$$\dot z = \operatorname{diag}[p_1(t), \dots, p_n(t)]z = P(t)z, \tag{5.3.16}$$

*where*

$$p_i = \frac{d}{dt} \ln \|x_i\|, \qquad i = 1, \dots, n,$$

*and the diagonal is integrally separated.*

PROOF. Take an integrally separated basis

$$X(t) = \{x_1(t), \ldots, x_n(t)\}$$

of system (5.0.1) and, according to it (similarly to Theorem 3.3.3), construct a Lyapunov matrix $L(t)$ realizing the transformation to a diagonal system,

$$L(t) = \left\{ \frac{x_1}{\|x_1\|}, \frac{x_2}{\|x_2\|}, \ldots, \frac{x_n}{\|x_n\|} \right\}.$$

Then

$$X(t) = L(t)\,\mathrm{diag}[\|x_1(t)\|, \ldots, \|x_n(t)\|]. \tag{5.3.17}$$

Consider the basis $Z(t)$ of system (5.3.16) such that

$$X(t) = L(t)Z(t);$$

this, according to (5.3.17), implies

$$Z(t) = \mathrm{diag}[\|x_1(t)\|, \ldots, \|x_n(t)\|].$$

At the same time,

$$P(t) \equiv \dot{Z}Z^{-1} = \mathrm{diag}\left[\frac{d}{dt}\ln\|x_1\|, \ldots, \frac{d}{dt}\ln\|x_n\|\right].$$

The separateness of the diagonal follows from (5.3.5). □

Without giving the proof, we present Millionshchikov's criterion **[28]** for a small change of directions of solutions of a linear system under small perturbations of the coefficients.

THEOREM 5.3.2 (Millionshchikov). *The following two statements are equivalent*:

1) *system* (5.0.1) *satisfies the condition* (5.3.3) *for integral separateness,*

2) *for any* $\varepsilon > 0$ *there exists a* $\delta > 0$ *such that if* $\sup_{t\in\mathbb{R}_+}\|Q(t)\| < \delta$, *then for any solution* $y(t)$ *of system* (5.0.4) *there exists a solution* $x(t)$ *of system* (5.0.1) *such that*

$$\sup_{t\in\mathbb{R}_+} \sphericalangle(x(t), y(t)) < \varepsilon,$$

*where the angle between the vectors* $x$ *and* $y$ *is taken in absolute value.*

## §4. Necessary and sufficient conditions for stability of characteristic exponents

At the beginning of the section we deal with a special Lyapunov transformation that is known as $H$*-transformation* in the theory of differential equations **[9]**.

Let a scalar function $p(t)$ be continuous and bounded on $\mathbb{R}_+$.

DEFINITION 5.4.1. The function

$$p^H(t) = \frac{1}{H}\int_t^{t+H} p(\tau)\,d\tau \tag{5.4.1}$$

is called the *Steklov function* for $p(t)$ with step $H$ $(H > 0)$ or the *Steklov average*.

LEMMA 5.4.1. *Functions $p_1(t)$ and $p_2(t)$ are integrally separated if and only if for sufficiently large $H$ their Steklov functions are separated in the standard sense:*

$$p_2^H(t) - p_1^H(t) \geqslant a > 0, \qquad t \in \mathbb{R}_+. \tag{5.4.2}$$

PROOF. First let us show that for a function $p(t)$ such that

$$\|p(t)\| \leqslant M, \qquad t \in \mathbb{R}_+,$$

the following equality holds:

$$\int_s^t p^H(x)\,dx = \int_s^t p(x)\,dx + I(t) - I(s), \tag{5.4.3}$$

$$|I(t) - I(s)| \leqslant MH, \qquad t \geqslant s \geqslant 0, \tag{5.4.4}$$

where

$$I(t) = \frac{1}{H}\int_t^{t+H} p(y)\,dy \int_{y-H}^t dx.$$

Indeed, from (5.4.1) we have

$$\begin{aligned}
&\int_s^t p^H(x)\,dx \\
&= \frac{1}{H}\int_s^t dx \int_x^{x+H} p(y)\,dy \\
&= \frac{1}{H}\left[\int_s^t p(y)\,dy \int_s^y dx + \int_t^{s+H} p(y)\,dy \int_s^t dx + \int_{s+H}^{t+H} p(y)\,dy \int_{y-H}^t dx\right] \\
&= \frac{1}{H}\int_s^t p(y)\,dy \int_{y-H}^y dx + \frac{1}{H}\left[-\int_s^t p(y)\,dy \int_{y-H}^s dx + \int_t^{s+H} p(y)\,dy \int_s^t dx \right.\\
&\qquad\qquad \left. + \int_{s+H}^{t+H} p(y)\,dy \int_{y-H}^t dx\right] \\
&= \int_s^t p(y)\,dy + \frac{1}{H}\left[-\int_s^t p(y)\,dy \int_{y-H}^s dx \right.\\
&\qquad + \int_t^{s+H} p(y)\,dy \left[\int_s^{y-H} dx + \int_{y-H}^t dx\right] \\
&\qquad\qquad \left. + \int_{s+H}^{t+H} p(y)\,dy \int_{y-H}^t dx\right] \\
&= \int_s^t p(y)\,dy + \frac{1}{H}\left[\int_t^{t+H} p(y)\,dy \int_{y-H}^t dx - \int_s^{s+H} p(y)\,dy \int_{y-H}^s dx\right].
\end{aligned}$$

By a straightforward calculation, we obtain

$$\frac{1}{H}\left|\int_t^{T+H} p(y)\,dy \int_{y-H}^t dx\right| \leqslant \frac{M}{H}\int_t^{t+H} (t - y + H)\,dy \leqslant \frac{MH}{2};$$

this implies the estimate (5.4.4).

Now let the condition (5.4.2) hold. Using (5.4.3) and (5.4.4), we have

$$\int_s^t (p_2(\tau) - p_1(\tau))\,d\tau = \int_s^t (p_2^H(\tau) - p_1^H(\tau))\,d\tau - I_2(t) + I_2(s) - I_1(s) + I_1(t)$$
$$\geqslant a(t-s) - 2MH;$$

this implies that the functions $p_1(t)$ and $p_2(t)$ are integrally separated.

Let, conversely, the functions $p_1(t)$ and $p_2(t)$ be integrally separated, i.e.,

$$\int_s^t (p_2(u) - p_1(u))\,du \geqslant a(t-s) - d, \qquad t \geqslant s \geqslant 0, \quad a > 0, \quad d \geqslant 0.$$

Consider the difference of Steklov's averages and estimate it from below,

$$p_2^H(t) - p_1^H(t) = \frac{1}{h}\int_t^{t+H} (p_2(\tau) - p_1(\tau))\,d\tau \geqslant a - \frac{d}{H},$$

i.e., (5.4.2) holds. □

DEFINITION 5.4.2 ($H$-transformation). Let the diagonal of the matrix $A(t)$ of system (5.0.1) be real; denote it by $A_d(t)$. Choose $H > 0$ and form the matrix

$$A_d^H(t) = \frac{1}{H}\int_t^{t+H} A_d(\tau)\,d\tau.$$

The transformation

$$x = \exp\int_0^t (A_d(\tau) - A_d^H(\tau))\,d\tau \cdot y \tag{5.4.5}$$

is said to be an *$H$-transformation*.

This transformation is Lyapunov, since the matrix of the transformation is bounded together with the inverse and its derivative. Indeed, the matrices $A_d(t)$ and $A_d^H(t)$ are bounded by the assumption (5.0.2); the boundedness of the matrix $\int_0^t (A_d - A_d^H)\,d\tau$ follows from (5.4.3).

THEOREM 5.4.1. *A diagonal real system with an integrally separated diagonal is reducible to a diagonal system with a separated diagonal.*

PROOF. Let the diagonal coefficients of the system

$$\dot{x} = \mathrm{diag}[a_1(t), \dots, a_n(t)]x = A(t)x$$

be integrally separated, i.e., let the inequalities (5.3.2) be valid. The transformation (5.4.5) reduces this system to the diagonal system

$$\dot{y}(L^{-1}AL - L^{-1}\dot{L})y = (A(t) - (A(t) - A^H(t)))y = A^H(t)y.$$

By Lemma 5.4.1, the diagonal of this system is separated in the standard sense. □

EXAMPLE 5.4.1. The system

$$\dot{x}_1 = (\cos t)x_1,$$
$$\dot{x}_2 = x_2$$

has an integrally separated diagonal which is not separated in the standard sense (see Example 5.3.2).

Using the $H$-transformation, we obtain the system

$$\dot{y}_1 = \frac{1}{H}(\sin(t+H) - \sin t)y_1,$$
$$\dot{y}_2 = y_2$$

with the diagonal separated in the standard sense for sufficiently large $H$.

We introduce the notion of growth of a vector-function to make the exposition more compact and explicit.

DEFINITION 5.4.3. Let a scalar function $r(t)$ be integrable on each finite interval of $\mathbb{R}_+$ and a vector-function $x(t)$ be such that

$$x(t) = O\left(\exp \int_0^t r(\tau)\,d\tau\right),$$

or

$$\|x(t)\| \leqslant D \exp \int_0^t r(\tau)\,d\tau,$$

where $D$ is a constant. In this case we say that the *growth of* $x(t)$ *is no higher than* $r$ and denote this by $x \prec r$.

THEOREM 5.4.2 [9]. *Let the functions* $p_1(t)$ *and* $p_2(t)$ *be separated,*

$$p_2(t) - p_1(t) \geqslant a > 0, \qquad t \in \mathbb{R}_+; \tag{5.4.6}$$

*then for any* $\varepsilon > 0$ $(0 < \varepsilon < a/2)$ *there exists a* $\delta > 0$ *such that the system*

$$\dot{x} = \operatorname{diag}[p_1(t), p_2(t)]x + Q(t)x = [P(t) + Q(t)]x \tag{5.4.7}$$

*for* $\sup_{t\in\mathbb{R}_+} \|Q(t)\| < \delta$ *has the following property: there are no solutions of growth intermediate between* $p_1 + \varepsilon$ *and* $p_2 - \varepsilon$, *i.e., any solution* $x(t)$ *of growth* $\prec p_2 - \varepsilon$ *is, in fact, of growth* $\prec p_1 + \varepsilon$ *and, moreover, these solutions admit the estimate*

$$\|x(t)\| \leqslant \|x(0)\| D_\varepsilon \exp \int_0^t (p_1(\tau) + \varepsilon)\,d\tau \tag{5.4.8}$$

*uniform in all such solutions and* $t \geqslant 0$.

PROOF. By the method of variation of parameters, any solution $x(t)$ of system (5.4.7) satisfies the integral equation

$$x(t) = \exp \int_0^t P(\tau)\,d\tau \left[x(0) + \int_0^t \exp\left(-\int_0^s P(\tau)\,d\tau\right) Q(s)x(s)\,ds\right].$$

Note that the fact that the unperturbed system is diagonal allows us to write the fundamental matrix in the form $\exp \int_0^t P(\tau)\, d\tau$. The components of the integral equation have the form

$$
(5.4.9)\qquad
\begin{aligned}
x_1(t) &= \exp \int_0^t p_1\, d\tau \left[ x_1(0) + \int_0^s \exp\left( -\int_0^t p_1\, d\tau \right) [Q(s)x(s)]_1\, ds \right], \\
x_2(t) &= \exp \int_0^t p_2\, d\tau \left[ x_2(0) + \int_0^t \exp\left( -\int_0^s p_2\, d\tau \right) [Q(s)x(s)]_2\, ds \right].
\end{aligned}
$$

Fix $\varepsilon > 0$, $0 < \varepsilon < a/2$, and choose a function $r(t)$ such that

$$
(5.4.10)\qquad p_1(t) + \varepsilon \leqslant r(t) \leqslant p_2(t) - \varepsilon, \qquad t \in \mathbb{R}_+.
$$

This is possible by the condition (5.4.6). Now let a solution $x(t)$ of growth $\prec r$ satisfy system (5.4.9). Then we have

$$
\left| \exp\left( -\int_0^s p_2\, d\tau \right) [Q(s)x(s)]_2 \right| \leqslant \delta D \exp\left( -\int_0^s p_2\, d\tau \right) \exp \int_0^s r\, d\tau \leqslant \delta D e^{-\varepsilon s}.
$$

The integral

$$
\int_0^\infty e^{-\int_0^s p_2\, d\tau} [Q(s)x(s)]_2\, ds
$$

converges; therefore, we must set $-x_2(0)$ equal to the value of this integral. Why? If the expression in the square brackets in the second equation (5.4.9) had a nonzero limit as $t \to \infty$, then the function $x_2(t)$ would be of growth $\prec p_2$, which contradicts our assumption. Therefore,

$$
x_2(t) = -e^{\int_0^t p_2\, d\tau} \int_t^\infty e^{-\int_0^s p_2\, d\tau} [Q(s)x(s)]_2\, ds.
$$

Taking into account the last correction, we again write system (5.4.9) in the form of one equation. To this end we introduce the matrix

$$
Z(t,s) = \begin{pmatrix} v(t,s) & 0 \\ 0 & w(t,s) \end{pmatrix},
$$

where

$$
\begin{aligned}
v(t,s) &= \begin{cases} \exp \int_s^t p_1(\tau)\, d\tau, & \text{for } s \leqslant t, \\ 0, & \text{for } s > t, \end{cases} \\
w(t,s) &= \begin{cases} -\exp \int_s^t p_2(\tau)\, d\tau, & \text{for } s \geqslant t, \\ 0, & \text{for } s < t. \end{cases}
\end{aligned}
$$

We set $x(0) = (x_1(0), 0)^\top$; then system (5.4.9) is written in the form

$$
(5.4.11)\qquad x(t) = e^{\int_0^t p_1\, d\tau} x(0) + \int_0^\infty Z(t,s)Q(s)x(s)\, ds.
$$

Let us study this system. Let $B^2(r)$ be the set of all continuous vector-functions $x(t)$ on $\mathbb{R}_+$ with values in $\mathbb{R}^2$ satisfying the estimate

$$
\|x(t)\| \leqslant D \exp \int_0^t r(u)\, du
$$

with some constant $D$, i.e., functions whose growth is $\prec r$. Obviously, this set represents a linear space, and for $r_1(t) < r_2(t)$, $t \in \mathbb{R}_+$, we have

$$B^2(r_1) \subseteq B^2(r_2). \tag{5.4.12}$$

We introduce the norm $\|x\|_r$ in $B^2(r)$ in the following way:

$$\|x\|_r = \sup_{t \in \mathbb{R}_+} \|x(t)\| e^{-\int_0^t r\,d\tau}. \tag{5.4.13}$$

The space $B^2(r)$ is isometric to the space of continuous bounded vector-functions

$$y(t) = x(t) \exp\left(-\int_0^t r(\tau)\,d\tau\right)$$

with the standard norm $\sup_{t \in \mathbb{R}_+} \|y(t)\|$. The convergence in this space means uniform convergence on $\mathbb{R}_+$. Hence, $B^2(r)$ is a complete linear normed space; from the convergence of the sequence $x_k(t)$ on $B^2(r)$ there follows uniform convergence on any finite interval of $\mathbb{R}_+$.

We write the equation (5.4.11) in the form

$$x(t) = \xi(t) + Jx(t). \tag{5.4.14}$$

By the definition of the matrix $Z(t,s)$ and the inequalities for $r(t)$ (see (5.4.10)), for any $t, s \in \mathbb{R}_+$ the estimate

$$\|Z(t,s)\| \leqslant \exp\left(\int_s^t r(u)\,du - \varepsilon|t-s|\right)$$

is valid. Indeed, for $t \geqslant s$ we have

$$\|Z(t,s)\| = \exp\left(\int_s^t p_1\,d\tau\right) \leqslant \exp\left(\int_s^t r(u)\,du - \varepsilon(t-s)\right);$$

for $t \leqslant s$

$$\begin{aligned}\|Z(t,s)\| &= \exp\int_s^t p_2\,d\tau = \exp\int_t^s -p_2(\tau)\,d\tau \\ &\leq \exp\int_t^s (-r(u)-\varepsilon)\,du = \exp\left(\int_s^t r(u)\,du - \varepsilon|t-s|\right).\end{aligned}$$

Further,

$$\begin{aligned}\|Jx(t) - Jy(t)\| &\leqslant \int_0^\infty \|Z(t,s)\| \cdot \|Q(t,s)\| \cdot \|x(s) - y(s)\|\,ds \\ &\leqslant e^{\int_0^t r\,d\tau} \int_0^t e^{-\varepsilon|t-s|} \|x-y\|_r \delta\,ds \leqslant e^{\int_0^t r\,d\tau} \frac{2\delta}{\varepsilon} \|x-y\|_r,\end{aligned}$$

or, for $\delta < \varepsilon/2$,

$$\|J_x - J_y\|_r \leqslant \theta \|x-y\|_r, \qquad \theta < 1. \tag{5.4.15}$$

Thus, the operator $J$ (see (5.4.14)) is a contraction and sends the space $B^2(r)$ into itself. Note that

$$\xi(t) = x(0) \exp\int_0^t p_1(\tau)\,d\tau;$$

therefore, $\xi \in B^2(r)$ and

$$\|\xi\|_r = \|x(0)\| = \|\xi(0)\|. \tag{5.4.16}$$

Indeed,

$$\left\| \xi(t) \exp\left( - \int_0^t r(u)\, du \right) \right\| \leqslant \|x(0)\|,$$

and the equality is attained at $t = 0$.

By the contraction mapping principle, the equation

$$x = \xi + Jx \tag{5.4.17}$$

has a unique solution $x$ in $B^2(r)$ and it can be obtained by the method of successive approximations,

$$x_0 = \xi, \qquad x_k = \xi + Jx_{k-1}.$$

Hence,

$$x = \lim_{k\to\infty} x_k = \xi + \sum_{i=0}^{\infty} (x_{i+1} - x_i).$$

By the linearity, we have

$$x = (I + J + J^2 + \cdots + J^k + \cdots)\xi.$$

Majorizing this series by using the estimates (5.4.15) for the solution of equation (5.4.17), we obtain

$$\|x\|_r \leqslant \frac{1}{1-\theta} \|\xi\|_r,$$

or, taking into account (5.4.16),

$$\|x(t)\| \leqslant \frac{\|x(0)\|}{1-\theta} e^{\int_0^t r(u)\, du}. \tag{5.4.18}$$

These arguments were carried out for an arbitrary function $r(t)$ satisfying the inequalities

$$p_1(t) + \varepsilon \leqslant r(t) \leqslant p_2(t) - \varepsilon, \qquad t \in \mathbb{R}_+.$$

With the decrease of $r(t)$ in these bounds, the space $B^2(r)$ only diminishes (see (5.4.12)), the fixed point is unique, and, therefore, is independent of the specific choice of $r(t)$.

This implies that the solutions of system (5.4.2) of growth $\prec p_2 - \varepsilon$ actually satisfy the estimate (5.4.18) for $r(t) = p_1(t) + \varepsilon$, i.e., the estimate (5.4.8) holds. □

We pass to necessary and sufficient conditions for the stability of the characteristic exponents of system (5.0.1). This result belongs to three authors: Bylov and Izobov (joint papers **[12, 13]**) and Millionshchikov **[31]**. These papers were published simultaneously in the same issue of the journal *Differential Equations* in 1969.

We give the proof of this result for two-dimensional systems, considering separately the cases of different and coinciding exponents. Then we formulate and discuss the criterion for the stability of exponents for systems of arbitrary order.

Consider a system

$$\dot{x} = A(t)x, \tag{5.4.19}$$

where $x \in \mathbb{R}^2$ and $A(t)$ is a $2 \times 2$ real matrix continuous and bounded on $\mathbb{R}_+$.

THEOREM 5.4.3. *If system (5.4.19) has different characteristic exponents* $\lambda_1 < \lambda_2$, *then their stability follows from the existence of a Lyapunov transformation* $x = L(t)z$ *of this system to a diagonal system*

$$\dot{z} = \operatorname{diag}[p_1(t), p_2(t)]z = P(t)z, \tag{5.4.20}$$

*where the functions* $p_1(t)$ *and* $p_2(t)$ *are integrally separated, i.e.,*

$$\int_s^t (p_2(\tau) - p_1(\tau))\,d\tau \geqslant a(t-s) - d, \qquad t \geqslant s \geqslant 0, \quad a > 0, \quad d \geqslant 0.$$

PROOF. Together with system (5.4.19), we consider the perturbed system

$$\dot{y} = [A(t) + Q(t)]y,$$

having the characteristic exponents $\lambda_1' \leqslant \lambda_2'$. We must show that for any $\varepsilon > 0$, there exists a $\delta > 0$ such that

$$|\lambda_i - \lambda_i'| < \varepsilon, \qquad i = 1, 2,$$

for $\sup_{t\in\mathbb{R}_+} \|Q(t)\| \leqslant \delta$. We assume from the beginning that the perturbed system is successively subjected to two Lyapunov transformations. The first is taken from the condition of the theorem (it reduces the matrix $A(t)$ to the diagonal one $P(t)$), the second is an $H$-transformation which changes the integral separateness of the functions $p_1(t)$ and $p_2(t)$ to the separateness in the standard sense. Recall that the Lyapunov transformations neither change the characteristic exponents themselves, nor affect their stability. The matrix $Q(t)$ under these transformations acquires four factors, i.e., two direct and two inverse matrices of these transformations, but the latter are bounded and the smallness of the perturbation is preserved.

Now we assume that the perturbed system has the form

$$\dot{y} = \operatorname{diag}[p_1(t), p_2(t)]y + Q(t)y = [P(t) + Q(t)]y, \tag{5.4.21}$$

where $\|Q(t)\| \leqslant \delta$ for $t \geqslant 0$, and

$$p_2(t) - p_1(t) \geqslant a > 0, \qquad t \in \mathbb{R}_+. \tag{5.4.22}$$

The fundamental matrix of the unperturbed system (5.4.20) has the form

$$X(t) = \operatorname{diag}\left[\exp\int_0^t p_1\,d\tau, \exp\int_0^t p_2\,d\tau\right],$$

and the characteristic exponent of any solution

$$x(t) = (x_1(t), x_2(t))^\top$$

such that $x_2(t) \neq 0$ is determined by the coordinate $x_2(t)$ (this can be seen from the inequality (5.4.22)); therefore, $x_2(t)$ determines the behavior of $\|x(t)\|$ as $t \to \infty$. This coordinate is said to be *leading*. From what follows we shall see that this behavior is preserved for the perturbed system as well.

Consider the solution

$$z(t) = (z_1(t), z_2(t))^\top$$

of system (5.4.21) such that $z_1(0) = \eta_1$, $z_2(0) = \eta_2$ and $|\eta_2| > |\eta_1|$. Multiplying the first eqation (5.4.21) by $z_1(t)$ and the second by $z_2(t)$, we obtain

$$z_i\dot{z}_i = p_i(t)z_i^2 + \sum_{k=1}^2 q_{ik}(t)z_i z_k, \qquad i = 1, 2,$$

or

$$\left|\frac{1}{2}\frac{dz_i^2}{dt} - p_i(t)z_i^2\right| \leqslant \sum_{k=1}^{2} |q_{ik}(t)z_i z_k|.$$

Hence, the following estimate holds:

$$-\delta\sum_{k=1}^{2}|z_i z_k| \leqslant \frac{1}{2}\frac{dz_i^2}{dt} - p_i(t)z_i^2 \leqslant \delta\sum_{k=1}^{2}|z_i z_k|. \tag{5.4.23}$$

Using the last inequality, we show that

$$|z_2(t)| > |z_1(t)| \qquad \text{for} \qquad t \geqslant 0. \tag{5.4.24}$$

This inequality, by virtue of the choice of initial data, is valid for $t = 0$ and, by continuity, in a neighborhood to the right of the origin. Let $t_2$ be the upper bound of $t$ such that (5.4.24) holds. Then

$$z_2^2(t_2) = z_1^2(t_2), \qquad z_2'(t_2) \leqslant z_1'(t_2). \tag{5.4.25}$$

The last inequality for the derivatives and the estimate (5.4.23) imply that

$$-2\delta z_2^2 + p_2(t_2)z_2^2 \leqslant \frac{1}{2}\frac{dz_1^2}{dt} \leqslant 2\delta z_2^2 + p_1(t_2)z_2^2.$$

We have $z_2(t_2) \neq 0$ since in the opposite case it would follow from (5.4.25) that

$$z_2(t_2) = z_1(t_2) = 0,$$

i.e., $z(t) \equiv 0$ for $t \geqslant 0$. Dividing the last inequality by $z_2^2$, for its extreme terms we write

$$-2\delta + p_2(t_2) \leqslant 2\delta + p_1(t_2),$$

or

$$p_2(t_2) - p_1(t_2) \leqslant 4\delta;$$

this contradicts (5.4.22), e.g., for $\delta \leqslant a/6$. Hence, (5.4.24) is proved.

The inequality (5.4.24) implies that

$$\chi[z_2] \geqslant \chi[z_1];$$

therefore,

$$\chi[z] = \chi[z_2].$$

We show that this exponent lies in a small neighborhood of $\lambda_2$. Since $z_2^2(t) \neq 0$ for $t \in \mathbb{R}_+$ (this follows from (5.4.24)), we divide the inequality (5.4.23) for $i = 2$ by $z_2^2(t)$. Then

$$p_2(t) - 2\delta \leqslant \frac{1}{2}\frac{d\ln z_2^2}{dt} \leqslant p_2(t) + 2\delta.$$

Let us integrate this inequality,

$$\int_0^t p_2(t)\,d\tau - 2\delta t \leqslant \ln|z_2(t)| - \ln|z_2(0)| \leqslant \int_0^t p_2(\tau)\,d\tau + 2\delta t. \tag{5.4.26}$$

Dividing the last inequality by $t$ and letting $t \to \infty$, we obtain

$$\lambda_2 - 2\delta \leqslant \chi[z_2] \leqslant \lambda_2 + 2\delta.$$

We have shown that there exists a characteristic exponent $\lambda_2'$ of system (5.4.21) such that $|\lambda_2 - \lambda_2'| < \varepsilon$ for $\delta < \varepsilon/2$.

We pass to the discussion of the second exponent of the perturbed system. We know its solution

$$z(t) = (z_1(t), z_2(t))^\top;$$

this allows us to lower the order of the system by the change

$$\begin{aligned} y_1 &= z_1(t) \int_a^t u(\tau)\, d\tau + v, \\ y_2 &= z_2(t) \int_a^t u(\tau)\, d\tau, \end{aligned} \tag{5.4.27}$$

where $a = 0$ if $\chi[u] \geqslant 0$, and $a = \infty$ if $\chi[u] < 0$, i.e., the integral is Lyapunov. Substituting (5.4.27) in system (5.4.21), we obtain

$$\dot{v} = p_1(t)v + q_{11}(t)v - z_1(t)u, \qquad z_2(t)u = q_{21}(t)v. \tag{5.4.28}$$

For the function $v(t)$, we have

$$\dot{v} = p_1(t)v + q_{11}(t)v - \frac{q_{21}(t)z_1(t)v}{z_2(t)},$$

or, by virtue of (5.4.28), we have

$$e^{\int_0^t p_1(\tau)\, d\tau - 2\delta t} v(0) \leqslant v(t) \leqslant v(0) e^{\int_0^t p_1(\tau)\, d\tau + 2\delta t},$$

or

$$\lambda_1 - 2\delta \leqslant \chi[v] \leqslant \lambda_1 + 2\delta. \tag{5.4.29}$$

The characteristic exponent of the solution $y(t)$ defined by the formula (5.4.27) coincides with the greatest of the exponents of its components. The inequality (5.4.24) implies

$$\chi\left[z_2 \int_a^t u\, d\tau\right] \geqslant \chi\left[z_1 \int_a^t u\, d\tau\right],$$

i.e., to obtain the final result it is necessary to compare the growth of $v(t)$ and $y_2(t)$. We estimate the exponent of $y_2(t)$, using (5.4.28) and the estimate (5.4.26):

$$\begin{aligned} |y_2(t)| &= \left| z_2(t) \int_a^t u\, d\tau \right| \leqslant |z_2(t)| \left| \int_a^t \left| \frac{q_{21} v}{z_2} \right| d\tau \right| \\ &\leqslant |\eta_2| e^{\int_0^t p_2\, d\tau + 2\delta t} \left| \int_a^t \frac{\delta \exp\left(\int_0^s p_1\, d\tau + 2\delta s\right)}{|\eta_2| \exp\left(\int_0^s p_2\, d\tau - 2\delta s\right)}\, ds \right| \\ &\leqslant e^{\int_0^t p_2\, d\tau + 2\delta t} \delta \left| \int_a^t e^{\int_0^s (p_1 - p_2)\, d\tau + 4\delta s}\, ds \right| \\ &\leqslant e^{\int_0^t p_2\, d\tau + 2\delta t} \delta \left| \int_a^t e^{-as + 4\delta s}\, ds \right|. \end{aligned}$$

Hence,

$$\chi[y_2] \leqslant \lambda_2 + 6\delta - a.$$

It follows from the last inequality that the exponent of the solution $y(t)$ is shifted to the right with respect to $\lambda_2$ by a finite distance, and the solution $y(t)$ has the growth

$\prec p_2 - a + 4\delta$. By Theorem 5.4.2, there are no solutions of growth between $p_1 + \varepsilon$ and $p_2 - \varepsilon$; therefore, the growth of $y(t) \prec p_1 + \varepsilon$, i.e.,

$$\lambda_1' = \chi[y] \leqslant \lambda_1 + \varepsilon.$$

It remains to show that $\lambda_1' \geqslant \lambda_1 - \varepsilon$. The solution $y(t)$ is defined by (5.4.27) and $\chi[y] \geqslant \chi[v]$. Indeed, the decrease of the exponent of the first component is possible only under the condition

$$\chi[v] = \chi\left[z_1 \int_a^t u\,d\tau\right],$$

but then $y_2(t)$ is the leading coordinate since

$$\chi[y_2] \geqslant \chi\left[z_1 \int_a^t u\,d\tau\right];$$

therefore,

$$\chi[y] = \chi[y_2] \geqslant \chi[v].$$

On the other hand, if

$$\chi\left[z_1 \int_a^t u\,d\tau\right] < \chi[v],$$

then

$$\chi[y] \geqslant \chi[y_1] = \chi[v].$$

Now the estimate for $\lambda_1'$ from below follows from (5.4.29) for $\delta < \varepsilon/2$. □

COROLLARY 5.4.1. *If the linear diagonal system* (5.4.20) *has different characteristic exponents* $\lambda_1 < \lambda_2$ *and its diagonal is separated, then the characteristic exponents are stable.*

COROLLARY 5.4.2. *If the linear diagonal system* (5.4.20) *has an integrally separated diagonal, then its characteristic exponents are stable and different.*

We give an example of a diagonal system of the second order with different but unstable characteristic exponents. We verify that its diagonal is not integrally separated.

EXAMPLE 5.4.2 (Perron).

$$\begin{aligned} \dot{x}_1 &= -ax_1, \\ \dot{x}_2 &= (\sin \ln t + \cos \ln t - 2a)x_2, \end{aligned} \qquad t \geqslant 0, \quad 1 > a > 1/2.$$

This system was considered in Example 4.4.1. Here

$$\lambda_1 = -a < \lambda_2 = 1 - 2a < 0.$$

For

$$1 < 2a < 1 + \frac{1}{2}\exp(-\pi), \tag{5.4.30}$$

the perturbation

$$Q(t) = \begin{pmatrix} 0 & 0 \\ e^{-at} & 0 \end{pmatrix}$$

generates a solution, unbounded as $t \to \infty$. This implies the instability of the exponents of the Perron system, since stable exponents do not change under the perturbations

$$\|Q(t)\| \to 0 \quad \text{as} \quad t \to \infty$$

(Theorem 5.2.1).

We show that the diagonal of the Perron system is not integrally separated if (5.4.30) holds. Indeed, in the opposite case, the inequality

$$\int_s^t (+a + \sin \ln \tau + \cos \ln \tau - 2a)\, d\tau \geqslant \alpha(t-s) - d,$$

i.e., the inequality

$$-a(t-s) + (t \sin \ln t - s \sin \ln s) \geqslant \alpha(t-s) - d \tag{5.4.31}$$

holds for all $t \geqslant s \geqslant 0$ with some constants $\alpha > 0$, $d \geqslant 0$. Take the sequences

$$t_m = \exp(2m+1)\pi, \qquad s_m = \exp(2m-1)\pi;$$

then from (5.4.31) for any $m \in \mathbb{N}$ we have

$$-ae^{2m\pi}(e^{\pi} - e^{-\pi}) \geqslant \alpha e^{2m\pi}(e^{\pi} - e^{-\pi}) - d.$$

Therefore, taking into account (5.4.30), we obtain

$$\begin{aligned} \alpha(e^{\pi} - e^{-\pi}) &\leqslant -a(e^{\pi} - e^{-\pi}) + de^{-2m\pi} \\ &< -\frac{1}{2}(e^{\pi} - e^{-\pi}) + de^{-2m\pi} \\ &< 0, \qquad m \to \infty; \end{aligned}$$

this contradicts (5.4.31). □

Actually, integral separateness of the diagonal is a necessary condition for the stability of characteristic exponents.

THEOREM 5.4.4 **[12]**. *If the exponents $\lambda_1 < \lambda_2$ of a two-dimensional system*

$$\begin{matrix} \dot{x}_1 = p_1(t)x_1, \\ \dot{x}_2 = p_2(t)x_2, \end{matrix} \qquad \frac{dx}{dt} = \operatorname{diag}[p_1(t), p_2(t)]x = P(t)x, \tag{5.4.32}$$

*are stable, then the functions $p_1(t)$ and $p_2(t)$ are integrally separated. Here*

$$\lambda_1 = \overline{\lim_{t\to\infty}} \frac{1}{t} \int_0^t p_1(\tau)\, d\tau,$$

$$\lambda_2 = \overline{\lim_{t\to\infty}} \frac{1}{t} \int_0^t p_2(\tau)\, d\tau.$$

PROOF. The idea of the proof is as follows. Assume that the diagonal is not integrally separated. Fixing $\beta \in (\lambda_1, \lambda_2)$, we show that under this assumption for any $\delta > 0$ the perturbation

$$\|Q(t)\| < \delta$$

can be chosen such that there exists a solution $y(t)$ of the perturbed system with $\chi[y] = \beta$, and this contradicts the stability of the exponents. The method for constructing these perturbations was developed by Millionshchikov **[28, 29, 23]** and is called the *method of rotations* or the *method of perturbations by rotations*.

Assume that the functions $p_1(t)$ and $p_2(t)$ are not integrally separated. Then for any $\delta > 0$ we can indicate an infinite sequence of intervals $\{[\tau_k, \theta_k]\}$ such that

$$d_k = \theta_k - \tau_k \to \infty, \qquad \tau_k \to \infty \qquad \text{monotonically as} \qquad k \to \infty,$$

and

$$\int_{\tau_k}^{\theta_k} (p_2 - p_1)\, d\tau < \frac{\delta}{4} d_k. \tag{5.4.33}$$

Fixing $\beta \in (\lambda_1, \lambda_2)$, we show that (5.4.33) allows us to indicate a matrix $Q(t)$, $\|Q(t)\| < \delta$, for any $\delta > 0$, so that the perturbed system

$$\frac{dy}{dt} = [P(t) + Q(t)]y \tag{5.4.34}$$

has a solution $y(t)$,

$$\chi[y] = \beta.$$

We choose $\delta > 0$ so that the inequalities

$$\lambda_1 + 3\delta/2 < \beta, \qquad \beta + \delta < \lambda_2 \tag{5.4.35}$$

are satisfied. First we use $\delta/2$-smallness of the perturbation in order to change the inequality (5.4.33), namely we consider the system

$$\begin{aligned} \dot{x}_1 &= (p_1 + \delta/2)x_1, \\ \dot{x}_2 &= p_2 x_2, \end{aligned} \qquad \dot{x} = P'(t)x, \tag{5.4.36}$$

with the exponents $\lambda_1 + \delta/2$, $\lambda_2$, which is intermediate between (5.4.32) and (5.4.34). In what follows, we shall deal with this system; its first coefficient is again denoted by $p_1(t)$, and the inequality (5.4.33) for this $p_1(t)$ is rewritten in the form

$$\int_{\tau_k}^{\theta_k} (p_2 - p_1)\, d\tau < -\frac{\delta}{4} d_k. \tag{5.4.37}$$

System (5.4.36) will be perturbed on some nonintersecting intervals $[t_0, t_0+1]$ by means of the method of rotations, carrying out in parallel the following two constructions:

1) of a perturbed system

$$\dot{y} = [P'(t) + Q(t)]y \tag{5.4.38}$$

with an already admissible perturbation $\|Q(t)\| < \delta/2$,

2) of its solution $y(t)$ so that $\chi[y]$ is equal to $\beta$.

The idea of the method of rotations consists in the following. In system (5.4.36), $x_2$ is the leading coordinate. Solutions with zero second coordinate have the growth $\prec p_1 + \delta/2$. If this solution is slightly rotated at a certain moment from the axis $x_1$ (i.e., the system is perturbed by a rotation), then it enters the sphere of influence of the coordinate $x_2$ and begins to accumulate the growth up to the level required. Then, by means of rotations, we bring the solution back to the axis $x_1$ and again rotate away from the axis at the next, determined in a special way, moment of time and keep it in the sphere of influence of $x_2$ until it accumulates the growth required, etc. Finally, we shall have constructed a solution

$$\|y(t)\| \leqslant \sin\omega \exp\beta t$$

and proved that there exists a sequence

$$t_k \to \infty$$

such that

$$\|y(t_k)\| = \sin\omega \exp \beta t_k.$$

This solves our problem.

Now what is $\omega$ and how is a rotation constructed? The solution $y(t)$ is "patched" from the solutions of system (5.4.36). The passage from one solution to another takes place on intervals $[t_0, t_0 + 1]$ of unit length and is carried out by an orthogonal transformation

$$y = U(t, \omega)x(t)$$

which results in a perturbation of system (5.4.36). Here

$$U(t, \omega) = \begin{pmatrix} \cos \pm\omega(t - t_0) & -\sin \pm\omega(t - t_0) \\ \sin \pm\omega(t - t_0) & \cos \pm\omega(t - t_0) \end{pmatrix}, \qquad 0 < \omega < \frac{\pi}{2},$$

$x(t)$ is the solution of system (5.4.36) such that $y(t_0) = x_0$; the sign before $\omega$ determines the direction of the rotation. For all $t \in [t_0, t_0 + 1]$ we have

$$\|x(t)\| = \|y(t)\|,$$

and, beginning with the moment $t_0 + 1$, we continue $y(t)$ along the solution $x(t)$ of system (5.4.36) such that

$$x(t_0 + 1) = y(t_0 + 1)$$

until the next rotation. The orthogonal transformation indicated leads to the system

$$\dot{y} = (UP'U^{-1} + \dot{U}U^{-1})y = [P'(t) + Q(t)]y.$$

Thus,

$$\begin{aligned} \|Q\| &= \|UP'U^{-1} + \dot{U}U^{-1} - P'\| \\ &= \|UP'U^{-1} - P'U^{-1} + P'U^{-1} - P'U^{-1}U + \dot{U}U^{-1}\| \\ &\leqslant \|(U - E)P'U^{-1} + P'U^{-1}(E - U) + \dot{U}U^{-1}\|. \end{aligned}$$

Note that $\|\dot{U}\| \leqslant \omega$; then, by the Lagrange formula, we have

$$\|U - E\| = \|U(t, \omega) - U(t_0, \omega)\| \leqslant \omega,$$

since $t - t_0 \leqslant 1$. We have $\|U^{-1}\| = 1$; hence,

$$\|Q(t)\| \leqslant (2\|P'\| + 1)\omega \leqslant (2M + 1)\omega, \tag{5.4.39}$$

if

$$|p_1(t)| \leqslant M, \qquad |p_2(t)| \leqslant M \qquad \text{for} \qquad t \in \mathbb{R}_+.$$

Choosing a number $\omega$ such that

$$\|Q(t)\| < \delta/2 \qquad \text{for} \qquad t \in \mathbb{R}_+,$$

i.e.,

$$\omega < \delta/[2(2M + 1)],$$

we say that the transformation is *admissible* (or *of admissible smallness*) if it is defined by a matrix $U(t, \omega_1)$, where $0 \leqslant \omega_1 \leqslant \omega$. In the passage from system (5.4.32) to system (5.4.38) the norm of the total perturbation is less than $\delta$, since it results from the perturbation of the coefficient $p_1(t)$ (of system (5.4.36)) and the perturbation by means of a rotation with the estimate (5.4.39). In what follows, when performing a concrete rotation, we do not write out $Q(t)$ explicitly, and only check that the angle of rotation is sufficiently small.

The construction of the solution $y(t)$ of system (5.4.38) will be carried out in cycles. Let us describe the first cycle in detail. Set $\theta_0 = 0$ and choose $T_0 > 1$ so that the following inequality holds:

$$\exp \int_0^t (p_1 + \delta)\, d\tau < \sin \omega \exp \beta t, \qquad t \geqslant T_0; \tag{5.4.40}$$

this is possible by virtue of (5.4.35). On the interval $[\theta_0, T_0 - 1]$ our $y(t)$ coincides with the solution $x(t)$ of system (5.4.36) defined by the initial conditions

$$x_1(0) = 1, \qquad x_2(0) = 0,$$

i.e., lies on the axis $x_1$ and is given by the formula

$$y(t) = \exp \int_0^t p_1(\tau)\, d\tau (1, 0)^\top,$$

and on the interval $[T_0 - 1, T_0]$ it is defined by the transformation

$$y(t) = U(t, \omega_1) \exp \int_0^t p_1\, d\tau (1, 0)^\top.$$

Here $0 \leqslant \omega_1 \leqslant \omega$, the rotation is carried out from the axis $x_1$ to the axis $x_2$, $\omega_1$ plays the key role in the first cycle and the choice of the required $\omega_1$ completes this cycle. Thus,

$$y(T_0) = \exp \int_{\theta_0}^{T_0} p_1(\tau)\, d\tau (\cos \omega_1, \sin \omega_1)^\top.$$

From the sequence of intervals $\{[\tau_k, \theta_k]\}$, which realize the inequality (5.4.37), we choose an interval (in what follows we assume that this is the first interval of the sequence and, thus, it has the form $[\tau_1, \theta_1]$) so that $\tau_1 > T_0 + 1$ and, moreover,

$$\sup_{T_0 \leqslant t \leqslant \tau_1} \frac{1}{t} \left[ \int_{\theta_0}^{T_0} p_1\, d\tau + \int_{T_0}^t p_2\, d\tau \right] > \beta + \delta; \tag{5.4.41}$$

this is possible by virtue of the inequalities (5.4.35) and (5.4.40). On the interval $[T_0, \tau_1 - 1]$, we continue $y(t)$ along the solution of system (5.4.36). On the interval $[\tau_1 - 1, \tau_1]$ this system is perturbed by means of a rotation, which we construct having previously defined the solution up to the point $\tau_1$ inclusive. Thus, $y(t)$ on the interval $[T_0, \tau_1]$ is defined by the vector

$$\begin{aligned} y(t) &= \operatorname{diag} \left[ \exp \int_{T_0}^t p_1\, d\tau, \exp \int_{T_0}^t p_2\, d\tau \right] y(T_0) \\ &= \exp \int_{\theta_0}^{T_0} p_1\, d\tau \left[ \cos \omega_1 \exp \int_{T_0}^t p_1\, d\tau, \sin \omega_1 \exp \int_{T_0}^t p_2\, d\tau \right]^\top. \end{aligned} \tag{5.4.42}$$

Two cases are possible:

1) the vector $y(\tau_1)$ forms an angle $\leqslant \omega$ with the axis $x_1$,
2) the vector $y(\tau_1)$ forms an angle $> \omega$ with the axis $x_1$.

In the first case, by means of a rotation on the interval $[\tau_1 - 1, \tau_1]$, we make the vector $y(t)$ at the moment $\tau_1$ lie on the axis $x_1$ and keep it there (as a solution of (5.4.36)) until the moment $\theta_1$, i.e., on the interval $[\tau_1, \theta_1]$ we have

$$y(t) = \|y(\tau_1)\| \exp \int_{\tau_1}^{t} p_1 \, d\tau (1, 0)^\top. \tag{5.4.43}$$

In the second case we cannot act in this way, since the angle of the rotation would have been greater than $\omega$ and we should have exceeded the admissible smallness of perturbations. In this case we carry out the transformation by means of rotation twice (on the intervals $[\tau_1 - 1, \tau_1]$ and $[\theta_1 - 1, \theta_1]$), and at the moment $\theta_1$ we make the vector $y(t)$ lie on the axis $x_1$. On the interval $[\tau_1 - 1, \tau_1]$ we rotate the vector $y(t)$ at the moment $\tau_1$ by the angle $\omega$ towards the axis $x_1$. The vector $y(\tau_1)$ defined by the formula (5.4.42) lies strictly inside the first quadrant since

$$\frac{y_2(\tau_1)}{y_1(\tau_1)} = \tan \omega_1 \exp \int_{T_0}^{\tau_1} (p_2 - p_1) \, d\tau < \infty;$$

therefore, after the rotation towards the axis $x_1$ on the interval $[\tau_1 - 1, \tau_1]$, the vector remains inside the first quadrant and forms an angle greater than $\omega$ with the axis $x_2$; hence,

$$\frac{y_2(\tau_1)}{y_1(\tau_1)} < \cot \omega. \tag{5.4.44}$$

On the interval $[\tau_1, \theta_1]$, we continue $y(t)$ as a solution of system (5.4.36) and show that at the moment $\theta_1$ the angle between the vector $y(\theta_1)$ and the axis $x_1$ is less than $\omega$. This enables us to carry out the transformation by means of rotation of admissible smallness on the interval $[\theta_1 - 1, \theta_1]$ and, at the moment $\theta_1$, to make $y(\theta_1)$ lie on the axis $x_1$. On the interval $[\tau_1, \theta_1]$, the vector $y(t)$ satisfies system (5.4.36); consequently,

$$y(t) = \left[ y_1(\tau_1) \exp \int_{\tau_1}^{t} p_1 \, d\tau, y_2(\tau_1) \exp \int_{\tau_1}^{t} p_2 \, d\tau \right]^\top.$$

By means of inequalities (5.4.44) and (5.4.37), we estimate the angle $\alpha$ between the vector $y(\theta_1)$ and the axis $x_1$:

$$\tan \alpha = \frac{y_2(\tau_1)}{y_1(\tau_1)} \exp \int_{\tau_1}^{\theta_1} (p_2 - p_1) \, d\tau < \cot \omega \exp \left( -\frac{1}{4} \delta d_1 \right).$$

The interval $[\tau_1, \theta_1]$ is chosen sufficiently far away to the right of $T_0$ so as to have the inequality (5.4.41); now we make this choice more precise assuming that $d_1$ is so large that

$$\cot \omega \exp(-\delta d_1 / 4) < \tan \omega.$$

This is possible because $d_k \to \infty$, $\tau_k \to \infty$ monotonically as $k \to \infty$. Under this choice we have $\tan \alpha < \tan \omega$, both angles are in the first quadrant; therefore, $0 < \alpha < \omega$. On the interval $[\theta_1 - 1, \theta_1]$, we carry out the transformation by means of rotation, which at the moment $\theta_1$ makes the vector $y(\theta_1)$ lie on the axis $x_1$, i.e.,

$$y(\theta_1) = \|y(\theta_1)\| [1, 0]^\top$$

((5.4.43) gives the same result). By construction, the vector-function $y(t)$ depends continuously on the parameter $\omega_1$, and, thus, the value of $\ln \|y(\theta_1)\|$ on the closed interval $0 \leqslant \omega_1 \leqslant \omega$ is finite.

Now we choose the number $T_1$ so that

$$\frac{1}{T_1}\left[\left(M+\frac{\delta}{2}\right)\theta_1+\sup_{0\leqslant\omega_1\leqslant\omega}\ln\|y(\theta_1)\|\right]<\delta. \tag{5.4.45}$$

We shall not perturb system (5.4.36) on the interval $[\theta_1, T_1]$, $y(t)$ is continued as a solution of this system. Therefore, for $\theta_1 \leqslant t \leqslant T_1$ we obtain

$$y(t)=\|y(\theta_1)\|\exp\int_{\theta_1}^{t}p_1\,d\tau(1,0)^{\top}.$$

Now, to complete the first cycle we have to discuss an important detail: the smallness of $\omega_1$. Consider the limiting cases.

1) $\omega_1 = 0$. In this case there were no perturbations by means of rotations at all, and on $[\theta_0, T_1]$ we have

$$y(t)=\exp\int_{0}^{t}p_1\,d\tau(1,0)^{\top};$$

by virtue of (5.4.40), for $T_0 \leqslant t \leqslant T_1$ the following inequality holds:

$$\|y(t)\|<\sin\omega\exp\beta t. \tag{5.4.46}$$

2) $\omega_1 = \omega$. On the interval $[T_0, \tau_1]$, by virtue of (5.4.42), we have

$$\|y(t)\|>y_2(t)=\sin\omega\exp\left[\int_{\theta_0}^{T_0}p_1\,d\tau+\int_{T_0}^{t}p_2\,d\tau\right].$$

The inequality (5.4.41) ensures the existense of a

$$t'\in[T_0,\tau_1]\subset[T_0,T_1]$$

such that

$$\|y(t')\|>\sin\omega\exp\beta t'. \tag{5.4.47}$$

Since the dependence of $\|y(t)\|$ on $\omega_1$ is continuous, by comparing the inequalities (5.4.46) and (5.4.47), we obtain the existence of an $\omega_1 \in (0, \omega)$ such that for the $y(t)$ constructed using this $\omega_1$ there exists a $t_1 \in [T_0, T_1]$ for which the following relations hold:

$$\begin{aligned}\|y(t)\|&\leqslant\sin\omega\exp\beta t, && t\subset[T_0,T_1],\\ \|y(t_1)\|&=\sin\omega\exp\beta t_1, && t_1\in[T_0,T_1].\end{aligned} \tag{5.4.48}$$

The first cycle of the construction of the vector-function $y(t)$, i.e., a solution of the perturbed system, is completed; we pass to the second cycle. We assume that the interval $[\tau_2, \theta_2]$ is located so far away from the point $T_1$ that

$$\sup_{T_1\leqslant t\leqslant\tau_2}\frac{1}{t}\left[\ln\|y(\theta_1)\|+\int_{\theta_1}^{T_1}p_1\,d\tau+\int_{T_1}^{t}p_2\,d\tau\right]>\beta+\delta. \tag{5.4.49}$$

This inequality is analogous to the inequality (5.4.41) of the first cycle. Further, we carry out the constructions according to the scheme of the first cycle replacing $\omega_1$ by $\omega_2$, and obtain the vector-function $y(t)$ on the interval $[\theta_2, T_2]$ defined by

$$y(t)=y(\theta_2)\exp\int_{\theta_2}^{t}p_1\,d\tau(1,0)^{\top}.$$

The number $T_2$ is chosen so that

$$\frac{1}{T_2}\left[\left(M+\frac{\delta}{2}\right)\theta_2+\sup_{0\leqslant\omega_2\leqslant\omega}\ln\|y(\theta_2)\|\right]<\delta.$$

This is an analog of the inequality (5.4.45) of the first cycle. Consider again the limiting cases for $\omega_2$.

1) $\omega_2=0$, i.e., no rotation on the interval $[T_1-1, T_1]$; then the equality

$$y(t)=\|y(\theta_1)\|\exp\left(-\int_0^{\theta_1}p_1\,d\tau\right)\exp\int_0^t p_1\,d\tau(1,0)^\top,\qquad t\in[T_1,T_2],$$

holds; thus, according to the choice of $T_0$ (see (5.4.40)) and $T_1$ (see (5.4.45)), we have the estimate

$$\|y(t)\|<\exp\int_0^t(p_1+\delta)\,d\tau<\sin\omega\exp\beta t,\qquad t\in[T_1,T_2].$$

2) $\omega_2=\omega$. In this case the perturbation by means of rotation has been carried out on the interval $[T_1-1, T_1]$; therefore,

$$y(T_1)=\|y(\theta_1)\|\exp\int_{\theta_1}^{T_1}p_1\,d\tau[\cos\omega_2,\sin\omega_2]^\top,$$

and for $t\in[T_1,\tau_2]\subset[T_1,T_2]$ we have

$$y(t)=\|y(\theta_1)\|\exp\int_{\theta_1}^{T_1}p_1\,d\tau\left[\cos\omega_2\int_{T_1}^t p_1\,d\tau,\sin\omega_2\int_{T_1}^t p_2\,d\tau\right]^\top;$$

hence,

$$\|y(t)\|\geqslant y_2(t)=\sin\omega\|y(\theta_1)\|\exp\left[\int_{\theta_1}^{T_1}p_1\,d\tau+\int_{T_1}^t p_2\,d\tau\right].$$

The inequality (5.4.49) guarantees the existence of a $t''\in[T_1,\tau_2]$ such that

$$\|y(t'')\|>\sin\omega\exp\beta t''.$$

Further, as in the first cycle, the continuous dependence of $y(t)$ on $\omega_2$ implies that there exist $\omega_2\in(0,\omega)$ and $t_2\in[T_1,T_2]$ such that the following relations (analogs of (5.4.48) of the first cycle) hold:

$$\begin{aligned}&\|y(t)\|\leqslant\sin\omega\exp\beta t &&\text{for all} && t\in[T_1,T_2],\\ &\|y(t_2)\|=\sin\omega\exp\beta t_2 &&\text{for some} && t_2\in[T_1,T_2].\end{aligned}$$

By induction, we can extend the construction of the vector-function $y(t)$ to the whole half-axis $\mathbb{R}_+$. As a result, we have a solution $y(t)$ of the perturbed system (5.4.38) with an admissible smallness of perturbation, and this solution has the following two properties: for all $t\in\mathbb{R}_+$ the estimate

$$\|y(t)\|\leqslant\sin\omega\exp\beta t$$

holds, and there exists a sequence $t_k\to\infty$, $k\to\infty$, such that

$$\|y(t_k)\|=\sin\omega\exp\beta t_k.$$

By definition, $\chi[y]=\beta$, and the theorem is proved. □

Combining Corollary 5.4.1 and Theorem 5.4.4, we obtain the following statement.

THEOREM 5.4.5. *If a linear two-dimensional diagonal system has different characteristic exponents, then they are stable if and only if the diagonal coefficients are integrally separated.*

Now, let us give one more example of a regular system having different and unstable characteristic exponents.

EXAMPLE 5.4.3.

$$\dot{x}_1 = 0,$$
$$\dot{x}_2 = \left(1 + \frac{\pi}{2}\sin \pi\sqrt{t}\right) x_2.$$

Here $\lambda_1 = 0$,

$$\begin{aligned}\lambda_2 &= \lim_{t\to\infty} \frac{1}{t}\int_0^t \left(1 + \frac{\pi}{2}\sin\pi\sqrt{\tau}\right) d\tau \\ &= \lim_{t\to\infty}\frac{1}{t}\left(t + \frac{1}{\pi}\sin\pi\sqrt{t} - \sqrt{t}\cos\pi\sqrt{t}\right) = 1.\end{aligned}$$

The system is regular by Lyapunov's Theorem 3.8.1. At the same time, the diagonal is not integrally separated. Indeed, let $n \in \mathbb{N}$, then

$$\int_{(2n-1)^2}^{(2n)^2} \left(1 + \frac{\pi}{2}\sin\pi\sqrt{\tau}\right) d\tau = 0;$$

therefore, we cannot indicate $a > 0$, $d \geqslant 0$ such that

$$\int_s^t \left(1 + \frac{\pi}{2}\sin\pi\sqrt{\tau}\right) d\tau \geqslant a(t-s) - d$$

would hold for all $t \geqslant s \geqslant 0$.

We pass to the case of a two-dimensional diagonal system with equal characteristic exponents. First, let us introduce a notation. Let $\{r(t)\}$ be the set of lower functions for system (5.0.1) (see Definition 5.1.2); then

$$\bar{\omega} = \sup_r \left\{ \overline{\lim_{t\to\infty}} \frac{1}{t}\int_0^t r(u)\,du \right\}.$$

The upper central exponent is denoted, as before, by $\Omega$.

THEOREM 5.4.6. *If a linear two-dimensional system has equal characteristic exponents $\lambda_1 = \lambda_2 = \lambda$, then they are stable if and only if*

$$\bar{\omega} = \lambda = \Omega.$$

PROOF. *Necessity.* Let the exponents of the system be stable but

$$\max\{\Omega - \lambda, \lambda - \bar{\omega}\} = \gamma > 0.$$

For the sake of definiteness, we assume that $\Omega - \lambda = \gamma > 0$. According to Vinograd's Theorem 5.1.5, there exists a perturbation $Q(t)$ of the initial system such that

$$\|Q(t)\| < \delta, \qquad t \geqslant 0,$$

where $\delta$ is arbitrarily small, and the perturbed system has a characteristic exponent $\lambda' \geqslant \Omega$. This contradicts the stability of the exponents. Similar arguments can be carried out for the case $\lambda - \bar{\omega} = \gamma > 0$. The fact that $\bar{\omega}$ is attainable was established by Millionshchikov [29].

*Sufficiency.* By Corollary 5.1.2, the condition $\lambda = \Omega$ guarantees that the exponent is rigid upwards. By analogous arguments [9], it can be shown that the inequality $\lambda = \bar{\omega}$ implies that the exponent is rigid downwards. □

The following three theorems are given without proof. Actually, all the arguments used in their proofs are similar to those in the proofs of Theorems 5.4.3, 5.4.4, 5.4.6 with the exception of the following one: the stability of exponents implies that there exists a Lyapunov transformation reducing the system to diagonal (different characteristic exponents) or block-triangular (the general case) form. The validity of this statement is again established by Millionshchikov's method of rotations. Recall that the block-triangular form of a system, as well as necessary and sufficient conditions for a system to be reducible to this form, were considered in Chapter III.

THEOREM 5.4.7 [**13**]. *If system* (5.0.1) *has different characteristic exponents* $\lambda_1 < \lambda_2 < \cdots < \lambda_n$, *then they are stable if and only if there exists a Lyapunov transformation* $x = L(t)z$ *of system* (5.0.1) *to the diagonal form*

$$\dot{z} = \operatorname{diag}[p_1(t), \dots, p_n(t)]z,$$

*where the functions*

$$p_i(t), \qquad p_{i+1}(t), \qquad i = 1, \dots, n-1,$$

*are integrally separated, i.e., for all* $t \geqslant s \geqslant 0$

$$\int_s^t (p_{i+1} - p_i)\, d\tau \geqslant a(t-s) - d, \qquad a > 0, \quad d \geqslant 0.$$

THEOREM 5.4.8 [**13**]. *If system* (5.0.1) *has different characteristic exponents*

$$\lambda_1 < \lambda_2 < \cdots < \lambda_n,$$

*then they are stable if and only if there exists a fundamental system of solutions*

$$X(t) = \{x_1(t), \dots, x_n(t)\}$$

*such that*

$$\frac{\|x_{i+1}(t)\|}{\|x_{i+1}(s)\|} : \frac{\|x_i(t)\|}{\|x_i(s)\|} \geqslant de^{\alpha(t-s)}, \qquad d \geqslant 1, \quad \alpha > 0.$$

THEOREM 5.4.9 [**13, 31**]. *The characteristic exponents of system* (5.0.1) *are stable if and only if there exists a Lyapunov transformation* $x = L(t)z$ *reducing system* (5.0.1) *to a block-triangular form*

$$\frac{dz}{dt} = \operatorname{diag}[P_1(t), P_2(t), \dots, P_q(t)]z,$$

*where each of the matrices* $P_k(t)$ *is upper-triangular of order* $n_k$ *and for the block systems*

$$\frac{dz^{(k)}}{dt} = P_k(t)z^{(k)}$$

*the following conditions hold*:

1) *all solutions of the block have the same characteristic exponent* $\Lambda_k$ *and*

$$\bar{\omega}_k = \Lambda_k = \Omega_k, \qquad k = 1, \dots, q,$$

2) *for any* $p_i(t)$ *from the diagonal of the block* $P_k(t)$ *and* $p_j(t)$ *from the diagonal of the block* $P_{k+1}(t)$ *the condition for integral separateness is satisfied, i.e., there exist constants* $a > 0$, $d \geqslant 0$ *such that for all* $t \geqslant s \geqslant 0$

$$\int_s^t (p_j - p_i)\, d\tau \geqslant a(t - s) - d, \qquad k = 1, 2, \dots, q - 1.$$

Let us discuss briefly the conditions of the theorem. When the characteristic exponents are stable, the linear space $L^n$ of solutions of system (5.0.1) can be represented as the direct sum of $q$ lineals $L_k$ of dimensions $n_k$, respectively, $\sum_1^q n_k = n$. All the solutions belonging to the lineal $L_k$ have the same exponent $\Lambda_k$. The angles

$$\alpha_k(t) = \sphericalangle\{M_k, L_{k+1}\},$$

where

$$M_k = L_1 \oplus L_2 \oplus \cdots \oplus L_k,$$

are such that

$$\inf_{t \in \mathbb{R}_+} \alpha_k(t) \geqslant \varepsilon > 0.$$

The upper central exponent $\Omega_k$ and the number $\bar{\omega}_k$ coincide with $\Lambda_k$; this guarantees that the exponent of the subspace is rigid upwards and downwards. Recall that Lyapunov transformations preserve central characteristic exponents. If the Cauchy matrix of the block

$$\dot{z}^{(k)} = P_k(t) z^{(k)}$$

is denoted by $U_k(t, \tau)$, then condition 2) of the theorem is equivalent to the following:

$$\| U_{k+1}^{-1}(t, \tau) \|^{-1} \geqslant d e^{a(t-\tau)} \| U_k(t, \tau) \|, \qquad t \geqslant \tau, \quad k = 1, \dots, q - 1.$$

In this case the blocks are said to be *integrally separated*. Integral separateness divides the spheres of influence of small perturbations of the system on the exponents of the subspaces and prevents the appearance of solutions of intermediate growth.

CHAPTER VI

# A Linear Homogeneous Equation of the Second Order

Consider the equation

$$\ddot{x} + a(t)\dot{x} + b(t)x = 0, \tag{6.0.1}$$

where $a(t)$ and $b(t)$ are real functions of the real argument $t$, continuous and bounded on some interval $I \subset \mathbb{R}$.

Equation (6.0.1) is the simplest equation that is not integrable explicitly. Unfortunately, there is no sufficiently complete correlation between the properties of solutions and the character of the coefficients of the equations. In this chapter we consider precisely this question and present some conditions on the coefficients that allow us to say whether the solutions of equation (6.0.1) are oscillating, bounded, and/or stable. We start with recalling some well-known facts **[5]**.

1. The solution of the Cauchy problem for equation (6.0.1) with the initial data $(t_0, x_0, x_0') \subset I \times \mathbb{R}^2$ exists, is unique, and is defined for all $t$ from $I$. In the general case, the interval $I$ is usually fixed by the choice of $t_0$. For example, for the equation

$$\ddot{x} + \frac{t}{t+1}\dot{x} + \frac{1}{t-1}x = 0$$

one of the three intervals

$$(-\infty, -1), \qquad (-1, 1), \qquad (1, \infty)$$

can be considered as $I$, depending on the position of $t_0$. Note that for some equations it may happen that one or both solutions constituting a fundamental system of solutions can be continued smoothly from one interval to another. For example, the equation $\ddot{x} + \dot{x} - x/t = 0$ has the solution $x = t$ defined on the whole real axis.

2. If a nontrivial solution $x_1(t)$ of equation (6.0.1) is known, then a linearly independent solution $x_2(t)$ is given by the formula

$$x_2(t) = x_1(t) \int_{t_0}^{t} \frac{\exp\left(-\int_{t_0}^{\tau} a(u)\,du\right)}{x_1^2(\tau)}\,d\tau, \qquad t_0, t \in I. \tag{6.0.2}$$

This result is obtained by reducing equation (6.0.1) by means of the substitution $x = x_1(t)\int_{t_0}^{t} z\,d\tau$ to a linear equation of the first order for the function $z$ and by integrating the latter. Note that if for some $\theta \in I$ we have $x_1(\theta) = 0$, then

$$\lim_{t \to \theta} x_2(t) = -\frac{\exp\left(\int_{t_0}^{\theta} a(u)\,du\right)}{\dot{x}_1(\theta)}$$

(this follows from (6.0.2) by the generalized L'Hospital rule **[34]**).

3. If the coefficient $a(t)$ is a continuously differentiable function on $I$, then the substitution

$$x = z \exp\left(-\frac{1}{2}\int_{t_0}^{t} a(\tau)\,d\tau\right), \qquad t_0, t \in I, \tag{6.0.3}$$

reduces equation (6.0.1) to the form

$$z'' + p(t)z = 0, \tag{6.0.4}$$

$$p(t) = -a^2(t)/4 - a'(t)/2 + b(t). \tag{6.0.5}$$

4. If two independent solutions of equation (6.0.1) are known, then the solution of the linear nonhomogeneous equation

$$\ddot{x} + a(t)\dot{x} + b(t)x = f(t), \qquad f \in C(I),$$

can be always obtained by the method of variation of parameters (the Lagrange method).

## §1. On the oscillation of solutions of a linear homogeneous equation of the second order

In this section we consider the question of zeros of nontrivial solutions of the equation

$$\ddot{x} + a(t)x + b(t)x = 0 \tag{6.1.1}$$

with continuous and bounded real coefficients on an interval $I \subset \mathbb{R}$.

DEFINITION 6.1.1. A solution $x(t)$ of equation (6.1.1) is said to be *oscillating* in the interval $(a, b) \subset I$ if it has at least two zeros in this interval. In the opposite case, $x(t)$ is a *nonoscillating* solution.

Thus, for example, all the solutions of the equation

$$\ddot{x} - m^2x = 0, \qquad m \in \mathbb{R}, \quad m \neq 0, \tag{6.1.2}$$

are nonoscillating on $\mathbb{R}$; this obviously follows from the form of the general solution

$$x = c_1 \exp mt + c_2 \exp(-mt).$$

Let us turn to the equation

$$\ddot{x} + m^2x = 0, \qquad m \in \mathbb{R}, \quad m \neq 0, \tag{6.1.3}$$

whose general solution

$$x = c_1 \cos mt + c_2 \sin mt$$

can be transformed to the form

$$x = A\sin(mt + \delta), \tag{6.1.4}$$

where $A = (c_1^2 + c_2^2)^{1/2}$ is the amplitude and

$$\delta = \arcsin\left(c_1/(c_1^2 + c_2^2)^{1/2}\right)$$

is the initial phase. It is clear from (6.1.4) that all the solutions of equation (6.1.3) have the period $2\pi/m$, the distance between two consecutive zeros of any solution is equal to $\pi/m$, and all the solutions are oscillating on any interval of length greater than $\pi/m$. Note that, with the increase of $m$, the distance between the zeros of the solutions decreases.

These two simple equations provide a rather accurate model of the oscillating character of solutions of the equation

$$\ddot{x} + p(t)x = 0; \tag{6.1.5}$$

the passage to which from the equation (6.1.1) is carried out by the substitution (6.0.3), preserving the zeros of the solutions. In what follows, we establish that for $p(t) \leqslant 0$ all the solutions of this equation are nonoscillating, while for $p(t) > 0$ they are oscillating and the frequency of oscillations grows with the growth of $p(t)$.

Before proving these results, we show that the question about the distance between consecutive zeros makes sense because of the following theorem.

LEMMA 6.1.1. *The zeros of any nontrivial solution $x(t)$ of equation (6.1.1) lying inside the interval $I$ are simple and isolated.*

PROOF. The first statement follows from the fact that only the trivial solution corresponds to the initial conditions $x(t_0) = 0$ and $\dot{x}(t_0) = 0$. Conversely, let there exist a limit point $t_0$ for the set $\{t_k\}$ of zeros of a nontrivial solution $x(t)$. By continuity, $x(t_0) = 0$. Let us show that then $\dot{x}(t_0) = 0$. Indeed,

$$\dot{x}(t_0) = \lim_{h \to 0} \frac{x(t_0 + h) - x(t_0)}{h}.$$

By choosing $h = t_k - t_0$, we obtain the required result. □

COROLLARY 6.1.1. *A nontrivial solution $x(t)$ of equation (6.1.1) on any compact from $I$ has a finite number of zeros.*

THEOREM 6.1.1 (Sturm). *Consider two equations*

$$y'' + p(t)y = 0 \qquad \textit{and} \qquad z'' + q(t)z = 0,$$

*where $p, q \in C(I)$ and*

$$q(t) \geqslant p(t), \qquad t \in I. \tag{6.1.6}$$

*Then between each two consecutive zeros of any nontrivial solution of the first equation there is at least one zero of any solution of the second equation under the condition that between these zeros there are points such that $q(t) > p(t)$.*

PROOF. Let $y_1(t)$ be the solution of the first equation such that

$$y_1(t_1) = y_1(t_2) = 0 \qquad \text{and} \qquad y_1(t) > 0 \qquad \text{for} \qquad t \in (t_1, t_2).$$

Assume that there exists a solution of the second equation $z_1(t)$ such that

$$z_1(t) > 0 \qquad \text{for} \qquad t \in (t_1, t_2)$$

(it is important for us to fix the sign of the solutions; this does not involve any loss of generality since $-y_1(t)$ and $-z_1(t)$ are also solutions with the same zeros). We substitute these solutions in the corresponding equations, multiply the first one by $z_1(t)$, the second by $y_1(t)$, and, subtracting the second identity from the first, we obtain

$$y_1'' z_1 - z_1'' y_1 = (q(t) - p(t)) z_1 \cdot y_1.$$

Let us integrate this identity from $t_1$ to $t_2$, taking into account that the left-hand side is the derivative of the difference $y_1'z_1 - y_1z_1'$ and that $y_1(t_1) = y_1(t_2) = 0$. Thus,

$$y_1'(t_2)z_1(t_2) - y_1'(t_1)z_1(t_1) = \int_{t_1}^{t_2} (q(\tau) - p(\tau))y_1z_1\, d\tau. \tag{6.1.7}$$

By the condition (6.1.6) of the theorem and our assumptions, the right-hand side of the identity (6.1.7) is strictly positive, and the left-hand side is nonpositive since

$$y_1'(t_2) > 0, \qquad z_1(t_2) \geqslant 0, \qquad y_1'(t_1) > 0, \quad z_1(t_1) \geqslant 0.$$

This contradiction is due to the assumption that

$$z_1(t) > 0 \qquad \text{for} \qquad t \in (t_1, t_2). \quad \square$$

COROLLARY 6.1.2. *The zeros of two linearly independent solutions of equation* (6.1.5) *mutually separate each other.*

PROOF. Let us denote by $y_1(t)$ and $z_1(t)$ linearly independent solutions and carry out the proof of Sturm's theorem up to the identity (6.1.7) inclusive. This time the right-hand side is equal to zero since $q(t) \equiv p(t)$, and the left-hand side is strictly negative since

$$z_1(t_2) > 0, \qquad z_1(t_1) > 0$$

(solutions constituting the fundamental system do not vanish simultaneously). Thus, $z_1(t)$ has only one zero inside $(t_1, t_2)$, since in the opposite case, changing the roles of the equations, we obtain one more zero of the solution $y_1(t)$ between $t_1, t_2$. $\square$

COROLLARY 6.1.3 (on the estimate of the distance between consecutive zeros of any nontrivial solution of equation (6.1.5)). *Let the inequality*

$$0 < l^2 \leqslant p(t) \leqslant L^2 \tag{6.1.8}$$

*be valid on a closed interval* $[\alpha, \beta] \subset I$. *Then the estimate*

$$\pi/L \leqslant d \leqslant \pi/l \tag{6.1.9}$$

*holds for the distance* $d$ *between consecutive zeros of any nontrivial solution of equation* (6.1.5).

PROOF. We apply successively Sturm's Theorem 6.1.1 to the pair of equations (6.1.5) and (6.1.3):

a) for $m = L$; this gives the left inequality in (6.1.9),

b) for $m = l$; this gives the right inequality in (6.1.9). $\square$

EXAMPLE 6.1.1. Consider the equation

$$\ddot{x} + 2t\dot{x} + (t^2 + t + 1)x = 0;$$

a) let us estimate the distance between consecutive zeros of any nontrivial solution on the interval **[16]**,

b) let us show that the distance between the zeros tends to zero as $t \to \infty$.

By the substitution (6.0.3)

$$x = ze^{-\frac{1}{2}\int_0^t 2\tau\, d\tau} = ze^{-t^2/2},$$

which does not move the zeros of solutions, we reduce the initial equation to the equation

$$\ddot{z} + tz = 0$$

(see (6.0.5)).

a) By the inequality $16 \leqslant t \leqslant 121$, according to Corollary 6.1.3 we have the estimate

$$\pi/11 \leqslant d \leqslant \pi/4.$$

b) Let $t \geqslant l^2$. Formula (6.1.9) implies that

$$0 < d \leqslant \pi/l \to 0 \qquad \text{as} \qquad l \to \infty.$$

COROLLARY 6.1.4. *If the inequality*

$$p(t) \geqslant l^2 > 0 \tag{6.1.10}$$

*holds for* $t \in I$, *then on any interval* $(\alpha, \beta) \subset I$ *such that* $\beta - \alpha > \pi/l$, *all the solutions of equation* (6.1.5) *are oscillating.*

Let $I = [\alpha, \infty)$. In this case the condition (6.1.10) can be relaxed.

THEOREM 6.1.2 [**4**]. *If*

$$p(t) \geqslant 0, \qquad \int_\alpha^\infty p(\tau)\,d\tau = \infty, \tag{6.1.11}$$

*then all the solutions of the equation*

$$\ddot{x} + p(t)x = 0$$

*have infinitely many zeros on the interval* $[\alpha, \infty)$.

PROOF. Let, to the contrary, there be a solution $x(t)$ preserving the sign beginning with some moment $t_0 \geqslant \alpha$; we assume that

$$x(t) > 0 \qquad \text{for} \qquad t \in [t_0, \infty).$$

From (6.1.5) we have

$$\ddot{x} = -p(t)x(t) \leqslant 0, \qquad t \geqslant t_0;$$

therefore, $x'(t)$ decreases monotonically. Two cases are possible:

1) $x'(t) > 0$ for $t \geqslant t_0$,

2) $x'(t) < 0$ for $t \geqslant t_1 \geqslant t_0$.

In the first case, the solution $x(t)$ increases monotonically and $x(t) \geqslant c$ for $t \geqslant t_0$. Hence,

$$\ddot{x} = -p(t)x(t) \leqslant -cp(t),$$

and, after integrating, we have

$$x'(t) - x'(t_0) \leqslant -c \int_{t_0}^t p(\tau)\,d\tau.$$

With the growth of $t$, the right-hand side of the last inequality tends to $-\infty$ by virtue of (6.1.11); this contradicts the condition

$$x'(t) > 0 \qquad \text{for} \qquad t \geqslant t_0.$$

We pass to the second case. Let $x'(t)$ be negative and decrease monotonically for $t \geqslant t_1$; consequently,

$$x'(t) \leqslant -c \qquad \text{for} \qquad t \geqslant t_1.$$

Integrating this inequality we obtain

$$x(t) - x(t_1) \leqslant -c(t - t_1),$$

which contradicts the condition that $x(t)$ is positive for all $t \geqslant t_0$. □

EXAMPLE 6.1.2. $\ddot{x} + \cos^2 tx = 0$.

The conditions of Theorem 6.1.2 are satisfied for this equation; therefore, all its solutions have infinitely many zeros on $\mathbb{R}$. Note that this result cannot be obtained by means of Sturm's theorem.

Consider one more result which allows one to obtain a lower bound for the distance between zeros. It is essential here that we use the form (6.1.1) for the equation.

THEOREM 6.1.3 (De la Vallée-Poussin [**35**]). *Let the coefficients of the equation*

$$\ddot{x} + a(t)\dot{x} + b(t)x = 0$$

*be such that*

$$|a(t)| \leqslant M_1, \qquad |b(t)| \leqslant M_2, \qquad t \in I. \tag{6.1.12}$$

*Then the estimates*

$$d \geqslant \frac{\sqrt{4M_1^2 + 8M_2} - 2M_1}{M_2}, \qquad M_2 > 0, \tag{6.1.13}$$

$$d \geqslant 2/M_1, \qquad M_2 = 0, \tag{6.1.14}$$

*are valid for the distance d between any two consecutive zeros of any nontrivial solution of equation* (6.1.1).

PROOF. Let $x(t)$ be a nontrivial solution of equation (6.1.1) and let $t_0$, $t_0 + d$ be its consecutive zeros. Without loss of generality we assume that $t_0 = 0$. Consider the identity

$$x'(t)d = \int_0^t \tau x''(\tau)\,d\tau - \int_t^d (d - \tau)x''(\tau)\,d\tau \tag{6.1.15}$$

which can easily be verified by integrating by parts and taking into account that $x(0) = x(d) = 0$. Let us change $x''(t)$ in (6.1.15) to its value from (6.1.1):

$$\begin{aligned} x'(t)d = &-\int_0^t \tau a(\tau)x'(\tau)\,d\tau + \int_t^d (d - \tau)a(\tau)x'(\tau)\,d\tau \\ &- \int_0^t \tau b(\tau)x(\tau)\,d\tau + \int_t^d (d - \tau)b(\tau)x(\tau)\,d\tau \end{aligned} \tag{6.1.16}$$

and let us estimate the right-hand side of (6.1.16).

Let

$$|x'(t)| \leqslant \mu \qquad \text{for} \qquad t \in [0, d];$$

$\mu > 0$ since $x(t)$ is a nontrivial solution. From the identities

$$x(t) = \int_0^t x'(\tau)\,d\tau \qquad \text{and} \qquad x(t) = \int_d^t x'(\tau)\,d\tau$$

we obtain the estimates

$$|x(t)| \leqslant \mu t, \qquad |x(t)| \leqslant \mu(d-t). \tag{6.1.17}$$

The first estimate will be used for $0 \leqslant t \leqslant d/2$, and the second for $d/2 \leqslant t \leqslant d$. Thus, (6.1.16) implies that

$$\begin{aligned} |x'(t)|d \leqslant M_1\mu &\left[\int_0^t \tau\,d\tau + \int_t^d (d-\tau)\,d\tau\right] \\ &+ M_2\left[\int_0^t \tau|x(\tau)|\,d\tau + \int_t^d (d-\tau)|x(\tau)|\,d\tau\right]. \end{aligned}$$

The first summand on the right-hand side of the last inequality is estimated by the number $M_1\mu d^2/2$; this is a straightforward result of the integration. To estimate the second summand, we divide the intervals of integration by the point $d/2$ and use the estimates (6.1.17), respectively. Finally, we have

$$|x'(t)|d \leqslant M_1\mu d^2/2 + M_2\mu d^3/8;$$

this holds for all $t \in [0, d]$ and, in particular, at the point where $|x'(t)| = \mu$. Hence, we obtain the inequality for $d$:

$$1 \leqslant M_1 d/2 + M_2 d^2/8. \tag{6.1.18}$$

The solution of this inequality implies the validity of the estimates (6.1.13) and (6.1.14). Recall that the number $d$, satisfying (6.1.18), lies to the right of the positive root of the equation

$$M_2 d^2 + 4M_1 d - 8 = 0. \quad \square$$

REMARK 6.1.1 (on the comparison of Theorems 6.1.1 and 6.1.3). When estimating the value of $d$ (the distance between consecutive zeros of solutions of the equation $\ddot{x} + a(t)\dot{x} + b(t)x = 0$) from below there appear two possibilities:

1) to use Theorem 6.3.1 straightforwardly,

2) to pass to the equation $\ddot{z} + p(t)z$ by means of the substitution (6.0.3), and, further, to carry out the estimate according to Corollary 6.1.3.

What is preferable? In the general case it is practically impossible to give a recommendation, but it should be noted that for sufficiently large $|a'(t)|$, $t \in I$, the second variant should be followed with caution. This results from the fact that

$$p(t) = -a^2(t)/4 - a'(t)/2 + b(t)$$

(see (6.0.4)), and an increase of $p(t)$ makes the left-hand side of the estimate (6.1.9) less accurate.

Consider some examples.

EXAMPLE 6.1.3. Let us estimate from below the distance between two neighboring zeros of nontrivial solutions of the following equations.

1) Consider Example 6.1.1:

$$\ddot{x} + 2t\dot{x} + (t^2 + t + 1)x = 0, \qquad 1 \leqslant t \leqslant 5.$$

a) By the Sturm theorem, we have

$$p(t) = t \quad \Rightarrow \quad 1 \leqslant p(t) \leqslant 5 \quad \Rightarrow \quad d > \pi/\sqrt{5} \approx 0.4.$$

b) By the de la Vallée-Poussin theorem (see (6.1.12) and (6.1.13)), we have

$$|a(t)| \leqslant 10 = M_1,$$

$$|b(t)| \leqslant 31 = M_2 \quad \Rightarrow \quad d \geqslant \frac{(4 \cdot 10 + 8 \cdot 31)^{1/2} - 2 \cdot 10}{31} \approx 0.2.$$

Thus, Sturm's theorem gives a better result

2) $\ddot{x} - (\arctan kt)\dot{x} + \pi^2 x = 0$, $t \in \mathbb{R}$.

a) By the Sturm theorem, we have

$$p(t) = -\frac{1}{4}(\arctan kt)^2 + \frac{k}{2(1 + k^2t^2)} + \pi^2 \leqslant \frac{k}{2} + \pi^2,$$

$$d > \frac{\pi\sqrt{2}}{(k + 2\pi^2)^{1/2}}.$$

b) By the de la Vallée-Poussin theorem, we have

$$|a(t)| \leqslant \frac{\pi}{2} = M_1, \qquad b(t) = \pi^2,$$

$$d \geqslant \frac{\left(\pi^2 + 8\pi^2\right)^{1/2} - \pi}{\pi^2} = \frac{2}{\pi}.$$

Under the condition $2k = \pi^4 - 4\pi^2$, the estimates coincide; for $2k > \pi^4 - \pi^2$ the second estimate is sharper.

We return to the equation (6.1.5) and consider the case when the coefficient $p(t)$ is nonpositive (see the model equation (6.1.2)).

THEOREM 6.1.4. *Let*

$$p(t) \leqslant 0, \qquad t \in I. \tag{6.1.19}$$

*Then all the nontrivial solutions of the equation*

$$\ddot{x} + p(t)x = 0$$

*are nonoscillating on $I$.*

PROOF. Let, to the contrary, there exist a solution $x(t) \not\equiv 0$ having two zeros, i.e.,

$$x(t_1) = x(t_2) = 0.$$

Let us appeal to Theorem 6.1.1. It implies that any solution of the equation $\ddot{z} = 0$ must intersect the axis $t$ in the interval $(t_1, t_2)$, which, obviously, is not true. □

The following theorem gives a necessary condition for equation (6.1.5) to have a solution with two zeros on $I$.

**Theorem 6.1.5** (Lyapunov [**37**]). *Let a solution $x \not\equiv 0$ of the equation*

$$\ddot{x} + p(t)x = 0$$

*have two zeros on the interval $[a, b] \subset I$. Then*

$$\int_a^b p^+(t)\,dt > \frac{4}{b-a}, \tag{6.1.20}$$

*where $p^+(t) = \max(p(t), 0)$.*

Proof. Consider the equation

$$u'' + p^+(t)u = 0, \tag{6.1.21}$$

which is majorant for (6.1.5). By Theorem 6.1.1, this equation has a solution $u(t)$ such that

$$u(\alpha) = u(\beta) = 0, \qquad a \leqslant \alpha < \beta \leqslant b,$$

and

$$u(t) > 0 \qquad \text{for} \qquad t \in (\alpha, \beta).$$

Consider the identity

$$-(\beta - \alpha)u(t) = (\beta - t)\int_\alpha^t (s-\alpha)u''(s)\,ds + (t-\alpha)\int_t^\beta (\beta - s)u''(s)\,ds. \tag{6.1.22}$$

Let us change the function $u''(s)$ in (6.1.22) to its expression from (6.1.21):
(6.1.23)

$$(\beta - \alpha)u(t) = (\beta - t)\int_\alpha^t (s-\alpha)p^+(s)u(s)\,ds + (t-\alpha)\int_t^\beta (\beta - s)p^+(s)u(s)\,ds.$$

Let

$$u(t_0) = \max u(t) \qquad \text{for} \qquad \alpha \leqslant t \leqslant \beta.$$

Set $t = t_0$ in (6.1.23), change $u(s)$ to $u(t_0)$ in the integrands (this leads to a strict inequality in (6.1.23)), and divide the result by $u(t_0)$. Hence,

$$\begin{aligned}\beta - \alpha &< (\beta - t_0)\int_\alpha^{t_0} (s-\alpha)p^+(s)\,ds + (t_0 - \alpha)\int_{t_0}^\beta (\beta - s)p^+(s)\,ds \\ &< \int_\alpha^{t_0} (\beta - s)(s-\alpha)p^+(s)\,ds \int_{t_0}^\beta (\beta - s)(s-\alpha)p^+(s)\,ds \\ &= \int_\alpha^\beta (\beta - s)(s-\alpha)p^+(s)\,ds,\end{aligned}$$

or

$$1 < \int_\alpha^\beta \frac{(\beta - s)(s-\alpha)}{\beta - \alpha}p^+(s)\,ds. \tag{6.1.24}$$

Note that the fraction in the integrand increases together with $\beta$ for $t \geqslant \alpha$ and decreases together with $\alpha$ for $t \leqslant \beta$; this is ensured by the signs of its derivatives with respect to $\beta$ and $\alpha$, correspondingly. By virtue of this fact, we have

$$1 < \int_a^b \frac{(b-s)(s-a)}{b-a}p^+(s)\,ds < \int_a^b \frac{(b-a)^2}{4(b-a)}p^+(s)\,ds;$$

this implies (6.1.20). In the last inequality we have used the fact that

$$4xy \leqslant (x+y)^2, \qquad x, y \in \mathbb{R}. \quad \square$$

COROLLARY 6.1.5. *If*

$$\int_a^b p^+(\tau)\,d\tau \leqslant \frac{4}{(b-a)} \tag{6.1.25}$$

*holds for* $[a,b] \subset I$*, then all the solutions of the equation*

$$\ddot{x} + p(t)x = 0$$

*are nonoscillating in* $[a,b]$.

Proof by contradiction.

EXAMPLE 6.1.4. Let us find the relation between the parameter $a > 0$ and the natural number $n$ such that all the solutions of the equation

$$\ddot{x} + xe^{-\alpha t}\sin t = 0$$

have no more than one zero on the interval $[0, 2n\pi]$.
We turn to the condition (6.1.25):

$$\begin{aligned}
\int_0^{2n\pi} [e^{-at}\sin t]^+\,dt &= \sum_{l=0}^{n-1}\int_{2l\pi}^{(2l+1)\pi} e^{-at}\sin t\,dt \\
&= \sum_{l=0}^{n-1}\frac{1}{a^2+1}\left(e^{-(2l+1)a\pi} + e^{-2la\pi}\right) \\
&= \frac{1}{a^2+1}\sum_{l=0}^{2n-1} e^{-al\pi} = \frac{1-e^{-2na\pi}}{(a^2+1)(1-e^{-a\pi})}.
\end{aligned}$$

Hence, the relation required has the form

$$\frac{1-e^{-2na\pi}}{(1+a^2)(1-e^{-a\pi})} \leqslant \frac{4}{2n\pi}.$$

COROLLARY 6.1.6 (on the number of zeros of a solution [37]). *Let* $N \geqslant 2$ *be the number of zeros of a nontrivial solution* $x(t)$ *of equation* (6.1.5) *on the interval* $0 \leqslant t \leqslant T$, $p \in C[0,T]$. *Then*

$$N < \frac{1}{2}\left(T\int_0^T p^+(s)\,ds\right)^{1/2} + 1. \tag{6.1.26}$$

Note that the choice of the interval is made for the sake of convenience and does not lessen generality. Let $x(t_i) = 0$ and $0 \leqslant t_1 < t_2 < \cdots < t_N \leqslant T$. By Theorem 6.1.5, we have

$$\int_{t_k}^{t_{k+1}} p^+(s)\,ds > \frac{4}{t_{k+1} - t_k}, \qquad k = 1, \ldots, N-1.$$

We sum these inequalities for $k = 1, \ldots, N-1$:

$$\int_{t_1}^{t_N} p^+(s)\,ds > 4 \sum_{k=1}^{N-1} \frac{1}{t_{k+1} - t_k}. \tag{6.1.27}$$

According to the inequality connecting the harmonic mean and the arithmetic mean of $N-1$ positive numbers, we have

$$\frac{1}{N-1} \sum_{k=1}^{N-1} \frac{1}{t_{k+1} - y_k} \geqslant \frac{N-1}{\sum_{k=1}^{N-1}(t_{k+1} - t_k)} = \frac{N-1}{t_N - t_1},$$

which, applied to (6.1.27), gives

$$\int_{t_1}^{t_N} p^+(s)\,ds > \frac{4(N-1)^2}{t_N - t_1},$$

or

$$\int_0^T p^+(s)\,ds > \frac{4(N-1)^2}{T}.$$

The last inequality implies (6.1.26). □

To conclude this section we note that some interesting criteria of nonoscillation for equation (6.0.5) are given in **[37]**, but in terms of solutions only.

## §2. On boundedness and stability of solutions of a linear equation of second order

The linear equation

$$\ddot{x} + a(t)\dot{x} + b(t)x = 0, \qquad a, b \in C(I \subset \mathbb{R}), \tag{6.2.1}$$

reduces to the linear system

$$\frac{dx}{dt} = \dot{x},$$
$$\frac{d\dot{x}}{dt} = -b(t)x - a(t)\dot{x};$$

in this interpretation all the results of the preceding chapters hold for this equation. Thus, equation (6.2.1) is Lyapunov stable, or, what is the same, all its solutions are stable, if and only if all the solutions of the equation and its derivatives are bounded as $t \to \infty$. Frequently, only the boundedness of solutions is of interest. In this section we consider some coefficient criteria for boundedness and stability of solutions of equation (6.2.1) which were proved, in particular, in **[4]** and **[35]**. Note that some interesting results of this sort were formulated in **[38]**.

Without loss of generality, we write equation (6.2.1) in the form

$$\ddot{x} + p(t)x = 0, \qquad p \in C(I \subset \mathbb{R}). \tag{6.2.2}$$

COMPARISON THEOREM 6.2.1 **[35]**. *Let there be given two equations*

$$y'' + p(t)y = 0, \qquad z'' + q(t)z = 0, \qquad p, q, \in C(I \subset \mathbb{R}),$$

*such that*

(6.2.3) 1) $q(t) \geqslant p(t), \qquad q(t) \not\equiv p(t), \qquad t \in I,$

2) $y(t)$ *and* $z(t)$ *are solutions of these equations such that*

(6.2.4) $$y(t_0) = z(t_0) = y_0, \qquad y'(t_0) = z'(t_0) = y_0', \quad t_0 \in I.$$

*Then, in any neighborhood to the right of* $t_0$ *in which* $z(t)$ *has no zeros and* $p(t) \not\equiv q(t)$, *the following inequality holds*:

(6.2.5) $$|y(t)| > |z(t)|;$$

*moreover, the ratio* $y(t)/z(t)$ *increases monotonically.*

PROOF. Denote by $(t_0, \alpha)$ the interval where $z(t) \neq 0$ and $p(t) \not\equiv q(t)$. For $z(t_0) \neq 0$ such an interval exists by the continuity of $z(t)$, and in the case $z(t_0) = 0$ it exists by virtue of $z'(t_0) \neq 0$ since we deal with a nontrivial solution. Substitute the solutions $y(t)$ and $z(t)$ in the corresponding equations; multiplying the first identity by $z(t)$, the second by $y(t)$, and subtracting the second from the first, we obtain

$$z(t)y''(t) - z''(t)y(t) = (q(t) - p(t))y(t)z(t), \qquad t \in (t_0, \alpha).$$

Integrating the result from $t_0$ to $t \in (t_0, \alpha)$ and taking into account the conditions (6.2.4), we obtain

(6.2.6) $$z(t)y'(t) - z'(t)y(t) = \int_{t_0}^{t} (q(\tau) - p(\tau))y(\tau)z(\tau)\, d\tau.$$

The integrand is positive for $t \in (t_0, \alpha)$ by virtue of (6.2.3) and the fact that $y(t)$ and $z(t)$ have the same sign. That this fact is true in a small neighborhood of $t_0$ follows from the conditions (6.2.4), and for other $t \in (t_0, \alpha)$ from the fact that $y(t)$ cannot vanish before $z(t)$. Indeed, let, to the contrary, there exist

$$t^* \in (t_0, \alpha) \qquad \text{such that} \qquad y(t^*) = 0.$$

For $t = t^*$, the right-hand side of (6.2.6) is strictly positive and the left-hand side is strictly negative, since the values $z(t^*)$ and $y'(t^*)$ necessarily have different signs. Thus, from (6.2.6) for $t \in (t_0, \alpha)$ we have

$$z^2(t)\frac{d}{dt}\left(\frac{y(t)}{z(t)}\right) > 0,$$

which proves the inequality (6.2.5). Note that $y(t_0)/z(t_0) = 1$; this is obvious for $y(t_0) \neq 0$ and for $y(t_0) = 0$ we have $y'(t_0) \neq 0$; then

$$\lim_{t \to t_0} \frac{y(t)}{z(t)} = \lim_{t \to t_0} \frac{y'(t)}{z'(t)} = 1. \qquad \square$$

EXAMPLE 6.2.1. Let us estimate the solution of the equation

$$\ddot{x} - (1 - (\cos^2 t)/2)x = 0, \qquad x(0) = 1, \quad x'(0) = 1,$$

on the interval $[0, \infty)$ from above and from below.

For the coefficient $p(t)$ we have the estimate $-1 \leqslant p(t) \leqslant -1/2$. Now we use Theorem 6.2.1. Consider $\ddot{x} - x = 0$ and $\ddot{x} - x/2 = 0$ as auxiliary equations with the initial conditions $x(0) = x'(0) = 1$.

Hence,

$$\frac{1-\sqrt{2}}{2}e^{-t/\sqrt{2}} + \frac{1+\sqrt{2}}{2}e^{t/\sqrt{2}} < x(t) < e^t.$$

The following theorem establishes the connection between the growth of the solutions of equation (6.2.2) with the monotonicity of the coefficient $p(t)$.

THEOREM 6.2.2 (Sonine-Polya **[35]**). *Let a function $p(t)$ be such that*
1) $p \in C^1(I), \quad p(t) \neq 0, \quad t \in I,$
2) $p'(t) \geqslant 0 \quad (p'(t) \leqslant 0), \quad t \in I.$

*Then for any solution $x(t)$ of equation* (6.2.2) *the values of $|x(t)|$ at the points of extremum form a nonincreasing (nondecreasing) sequence.*

PROOF. Take a nontrivial solution $x(t)$ of equation (6.2.2) and form the function

$$\varphi(t) = x^2(t) + \frac{1}{p(t)}(x'(t))^2.$$

By virtue of equation (6.2.2),

$$\varphi(t) = -\frac{1}{p^2(t)}(x'(t))^2 p'(t). \tag{6.2.7}$$

Note that at the points of extremum of the solution ($x'(t) = 0$) we have $\varphi(t) = x^2(t)$. The function $\varphi(t)$ is nonincreasing with respect to $t$ for $p'(t) \geqslant 0$, since, by virtue of (6.2.7), $\varphi'(t) \leqslant 0$, and $\varphi(t)$ is nondecreasing for $p'(t) \leqslant 0$; this implies our claim. □

COROLLARY 6.2.1. *If $p'(t) \geqslant 0$, $t \in I$, then for any solution $x(t)$ of equation* (6.2.2) *with initial conditions $t_0$, $x_0$, $x_0' = 0$ we have*

$$|x(t)| \leqslant |x_0| \qquad \textit{for} \qquad t \geqslant t_0.$$

PROOF. $|x(t_0)| = |x_0|$ is the first term of the sequence of values of $|x(t)|$ at its points of extremum. To finish the proof, one proceeds according to Theorem 6.2.2. □

EXAMPLE 6.2.2. The solution $x(t)$ of the equation $\ddot{x} + (1 + t^2)x = 0$ with the initial conditions $t_0 = 0$, $x_0 = 1$, $x_0' = 0$ satisfies the inequality $|x(t)| \leqslant 1$ for $t \geqslant 0$.

COROLLARY 6.2.2. *Let*

$$p \in C^1(I), \qquad p(t) < 0, \qquad p'(t) \geqslant 0 \qquad \textit{for} \qquad t \in I.$$

*Then any solution $x(t)$ of equation* (6.2.2) *with initial conditions $t_0$, $x_0$, $x_0'$ such that*

$$x_0^2 + \frac{1}{p(t_0)}(x_0')^2 \leqslant 0 \tag{6.2.8}$$

*is monotonic for $t \geqslant t_0$.*

PROOF. For the trivial solution the statement is obvious. Let $x(t)$ be a nontrivial solution; then (6.2.8) implies that $x_0' \neq 0$. By the definition of the function $\varphi(t)$, we have

$$\varphi(t_0) \leqslant 0, \qquad \varphi'(t_0) \leqslant 0, \qquad \varphi'(t) \leqslant 0;$$

hence,

$$\varphi(t) \leqslant 0 \qquad \text{for} \qquad t \geqslant t_0.$$

Therefore,

$$x^2(t) + \frac{1}{p(t)}(x'(t))^2 \leqslant 0 \qquad \text{for} \qquad t \geqslant t_0.$$

The last inequality implies that

$$x'(t) \neq 0, \qquad t \geqslant t_0. \quad \square$$

EXAMPLE 6.2.3. All the solutions $x(t)$ of the equation $\ddot{x} - (1 + 1/t)x = 0$ with initial conditions $t_0 \geqslant 1$, $x_0$, $x_0'$ such that

$$x_0^2(t_0 + 1) - t_0(x_0')^2 \leqslant 0$$

are monotonic.

The following theorem is valid for arbitrary functions $u(t)$ such that $u \in C^2(\mathbb{R}_+)$ and can be used in our problems.

THEOREM 6.2.3. *Let $u \in C^2(\mathbb{R}_+)$; then the boundedness of $|u(t)|$ and $|u''(t)|$ implies that $|u'(t)|$ is bounded for $t \in \mathbb{R}_+$.*

PROOF. Let

$$\sup_{t \in \mathbb{R}_+} \{|u(t)|, \quad |u''(t)|\} = M.$$

Let us set

(6.2.9) $$u'' - u = f(t);$$

considering $u(t)$ as a solution of equation (6.2.9), by the Lagrange method we obtain

$$u(t) = c_1 e^t + c_2 e^{-t} + \frac{e^t}{2} \int_0^t f(\tau) e^{-\tau}\, d\tau - \frac{e^{-t}}{2} \int_0^t f(\tau) e^{\tau}\, d\tau.$$

Since $|f(t)| \leqslant 2M$ for $t \in \mathbb{R}_+$, the last summand has a finite limit as $t \to \infty$ (L'Hospital's rule); therefore, it is bounded. The boundedness of $|u(t)|$ necessarily implies that

$$c_1 = -\frac{1}{2} \int_0^\infty f(\tau) e^{-\tau}\, d\tau.$$

Hence, we find that

$$u(t) = -\frac{e^t}{2} \int_t^\infty f(\tau) e^{-\tau}\, d\tau + c_2 e^{-t} - \frac{1}{2} e^{-t} \int_0^t f(\tau) e^{\tau}\, d\tau;$$

therefore,

(6.2.10) $$u'(t) = -\frac{e^t}{2} \int_t^\infty f(\tau) e^{-\tau}\, d\tau - c_2 e^{-t} + \frac{1}{2} e^{-t} \int_0^t f(\tau) e^{\tau}\, d\tau.$$

All three summands on the right are bounded for $t \in \mathbb{R}_+$; thus,

$$\sup_{t \geqslant 0} |u'(t)| < \infty. \quad \square$$

COROLLARY 6.2.3. *If* $\sup_{t \geqslant t_0} |p(t)| < \infty$ *in equation* (6.2.2), *then the boundedness of all its solutions implies their stability.*

COROLLARY 6.2.4. *Let* $|p(t)| \leqslant M$, $t \in \mathbb{R}_+$; *then, if all the solutions of equation* (6.2.2) *tend to zero, they are asymptotically stable.*

PROOF. Equation (6.2.2) is asymptotically stable if and only if all its solutions together with the derivatives tend to zero. Let $x(t) \to 0$ as $t \to \infty$; from equation (6.2.2) we have also that $x''(t) \to 0$ as $t \to \infty$, since $|p(t)| \leqslant M$. Set $f(t) = x''(t) - x(t)$; then equality (6.2.10), where $f(t) \to 0$, is valid for $x'(t)$; therefore,

$$x'(t) \to 0 \qquad \text{for} \qquad t \to \infty. \quad \square$$

Let us turn to the equation $\ddot{x} + m^2 x = 0$. It is stable since all its solutions are bounded together with their derivatives (see (6.1.4)). Consider the perturbed equation

$$\ddot{x} + [m^2 + \psi(t)]x = 0, \qquad m \neq 0, \tag{6.2.11}$$

and find out for which $\psi(t)$ the stability is preserved. As was shown in Theorem 4.3.1, the condition

$$\int_{t_0}^{\infty} |\psi(\tau)|\, d\tau < \infty$$

is sufficient, and it was noted in the same theorem that the stability of linear systems with constant coefficients may not be preserved under a perturbation which merely tends to zero as $t \to \infty$. To confirm this for equation (6.2.11), we consider an example.

EXAMPLE 6.2.4 [**4**]. The equation

$$\ddot{x} + \left[1 + \frac{4}{t}\cos t \sin t + \frac{1}{t^2}\cos^2 t \sin^2 t\right] x = 0$$

has the solution

$$x(t) = \cos t \exp \int_1^t \frac{\cos^2 \tau}{\tau}\, d\tau,$$

which is unbounded because the integral is unbounded as $t \to \infty$.

THEOREM 6.2.4 [**4**]. *Let*
1) $\psi(t) \to 0$ *as* $t \to \infty$,
2) $\int_{t_0}^{\infty} |\psi'(\tau)|\, d\tau < \infty$.
*Then all the solutions of equation* (6.2.11) *are stable.*

PROOF. Let us show that the solutions of equation (6.2.11) are bounded as $t \to \infty$. Take a nontrivial solution $x(t)$, substitute it in the equation, multiply the left-hand side by $x'(t)$, and integrate the result from $t_0$ to $t$:

$$\frac{1}{2}x'^2(t) + \frac{1}{2}x^2(t) + \int_{t_0}^{t} \psi(\tau)x(\tau)x'(\tau)\, d\tau = c_1, \qquad m = 1.$$

Further, integrating by parts, we find that

$$\frac{x'^2(t)}{2}+\frac{x^2(t)}{2}+\psi(t)\frac{x^2(t)}{2}-\frac{1}{2}\int_{t_0}^{t}\psi'(\tau)x^2(\tau)\,d\tau=c_2. \tag{6.2.12}$$

Since $\psi(t)\to 0$ as $t\to 0$, there exists a sufficiently large $t_0$ such that the inequality $1+\psi(t)\geqslant 1/2$ holds for $t\geqslant t_0$. From the identity (6.2.12) we have

$$x^2(t)\leqslant 4|c_2|+2\int_{t_0}^{t}|\psi'(t)|x^2(\tau)\,d\tau.$$

Applying the Gronwall-Bellman lemma (see Appendix) to the last inequality, we obtain

$$x^2(t)\leqslant 4|c_2|\exp\left(2\int_{t_0}^{t}|\psi'(\tau)|\,d\tau\right);$$

this ensures the boundedness of $x(t)$ as $t\to\infty$. Thus, all the solutions of equation (6.2.11) are bounded; this, according to Corollary 6.2.3, implies their stability. □

REMARK 6.2.1. It is clear from Example 6.2.4 that the second condition of the theorem cannot be changed to $\psi'(t)\to 0$ as $t\to\infty$.

## §3. Linear equations with periodic coefficients

In this section we consider coefficient criteria, due to Lyapunov, for stability and instability of the equation

$$\ddot{x}+p(t)x=0, \tag{6.3.1}$$

where

$$p(t+\omega)=p(t),\qquad p\in C(\mathbb{R}),$$

and give a generalization of the stability criterion.

Equation (6.3.1) is equivalent to the system

$$\begin{aligned}\frac{dx}{dt}&=\dot{x},\\ \frac{d\dot{x}}{dt}&=-p(t)x,\end{aligned} \tag{6.3.2}$$

whose stability is determined by the location of the multipliers with respect to the unit circle (Theorem 4.2.3).

Let $f(t)$ and $\varphi(t)$ be solutions of equation (6.3.1) such that

$$f(0)=1,\qquad f'(0)=0, \tag{6.3.3}$$

$$\varphi(0)=0,\qquad \varphi'(0)=1. \tag{6.3.4}$$

The matriciant of system (6.3.2) has the form

$$X(t,0)=\begin{pmatrix} f(t) & \varphi(t)\\ f'(t) & \varphi'(t)\end{pmatrix},$$

and the multipliers of the system are the eigenvalues of the matix $X(\omega,0)$. Note that

$$\operatorname{Det}X(\omega,0)=\operatorname{Det}X(0,0)\exp\int_0^{\omega}\operatorname{Sp}A(\tau)\,d\tau=1,$$

since the matrix of the coefficients of system (6.3.2) has trace zero and $X(0,0) = E$. We write the equation determining the multipliers in the form

$$\operatorname{Det}(X(\omega,0) - \rho E) \equiv \rho^2 - A\rho + 1 = 0, \tag{6.3.5}$$

$$A = \operatorname{Sp} X(\omega,0) = f(\omega) + \varphi'(\omega). \tag{6.3.6}$$

The number $A$ is called the *Lyapunov constant.* From (6.3.5) we have

$$\rho_{1,2} = (A \pm \sqrt{A^2 - 4})/2;$$

hence, by Theorem 4.2.3, we have the following statements.

PROPOSITION 6.3.1. *If $|A| > 2$, then both multipliers $\rho_1$ and $\rho_2$ are real, different, and $|\rho_1| > 1$, $|\rho_2| < 1$; therefore, system (6.3.2) and, consequently, also equation (6.3.1) are unstable.*

PROPOSITION 6.3.2. *If $|A| < 2$, then $\rho_1 = \bar{\rho}_2$, $|\rho_1| = |\rho_2| = 1$; therefore, system (6.3.2) and, consequently, also equation (6.3.1) are stable.*

PROPOSITION 6.3.3. *If $|A| = 2$, then $\rho_1 = \rho_2 = \rho$, $|\rho| = 1$ and, consequently, system (6.3.2) has a periodic ($\rho = 1$) or antiperiodic ($\rho = -1$) solution, and the problem of stability reduces to the study of the canonical structure of the matrix $X(\omega,0)$.*

PROPOSITION 6.3.4. *System (6.3.2) and, consequently, equation (6.3.1) cannot be asymptotically stable, since $\rho_1\rho_2 = 1$ and the multipliers cannot simultaneously lie strictly inside the unit circle.*

Thus, to determine the character of stability of equation (6.3.1), we have to estimate the Lyapunov constant $A$ defined by (6.3.6). To this end, we use the method of parameter.

We write equation (6.3.1) in the form

$$\ddot{x} = \varepsilon p(t)x. \tag{6.3.7}$$

For $\varepsilon = -1$ it coincides with equation (6.3.1). Let $f(t,\varepsilon)$ and $\varphi(t,\varepsilon)$ be the solutions of equation (6.3.7) whose initial data are consistent with (6.3.3) and (6.3.4), i.e.,

$$f(0,\varepsilon) = 1, \qquad f'(0,\varepsilon) = 0, \tag{6.3.8}$$

$$\varphi(0,\varepsilon) = 0, \qquad \varphi'(0,\varepsilon) = 1. \tag{6.3.9}$$

By the analyticity of the matriciant with respect to the parameter (Chapter I, §2, property 5), both solutions can be represented in the domain $t \in \mathbb{R}$, $|\varepsilon| < \infty$ by the series

$$f(t,\varepsilon) = \sum_{k=0}^{\infty} f_k(t)\varepsilon^k, \tag{6.3.10}$$

$$\varphi(t,\varepsilon) = \sum_{k=0}^{\infty} \varphi_k(t)\varepsilon^k, \tag{6.3.11}$$

whose coefficients are determined recurrently from equation (6.3.7) under the corresponding choice of initial conditions. An analogous statement also holds for the derivatives $\dot{f}(t,\varepsilon)$ and $\dot{\varphi}(t,\varepsilon)$.

We substitute the series (6.3.10) in equation (6.3.7), equate the coefficients of like powers of $\varepsilon$ on the left- and right-hand sides, and, taking into account (6.3.8), we obtain

$$\ddot{f}_0(t) = 0, \qquad f_0(0) = 1, \qquad \dot{f}_0(0) = 0,$$
$$\ddot{f}_k = p(t) f_{k-1}(t), \qquad f_k(0) = 0, \qquad \dot{f}_k(0) = 0, \qquad k = 1, 2, \dots,$$

or

$$\begin{aligned} f_0(t) &= 1, \\ f_k(t) &= \int_0^t dt_1 \int_0^{t_1} p(t_2) f_{k-1}(t_2)\, dt_2 \\ &= \int_0^t dt_2 \int_{t_2}^t p(t_2) f_{k-1}(t_2)\, dt_1 \\ &= \int_0^t (t - t_2) p(t_2) f_{k-1}(t_2)\, dt_2, \qquad k = 1, 2, \dots. \end{aligned}$$

Hence,
(6.3.12)

$$\begin{aligned} f(t) = f(t, -1) = 1 &- \int_0^t (t - t_1) p(t_1)\, dt_1 \\ &+ \int_0^t (t - t_1) p(t_1)\, dt_1 \int_0^{t_1} (t_1 - t_2) p(t_2)\, dt_2 + \cdots \\ &+ (-1)^k \int_0^t (t - t_1) p(t_1)\, dt_1 \cdots \int_0^{t_{k-1}} (t_{k-1} - t_k) p(t_k)\, dt_k + \cdots. \end{aligned}$$

Similarly, from (6.3.7), by virtue of (6.3.11) and the initial conditions (6.3.9), we have

$$\ddot{\varphi}_0(t) = 0, \qquad \varphi_0(0) = 0, \qquad \dot{\varphi}_0(0) = 1,$$
$$\ddot{\varphi}_k = p(t)\varphi_{k-1}(t), \qquad \varphi_k(0) = 0, \qquad \dot{\varphi}_k(0) = 0, \qquad k = 1, 2, \dots,$$

or

$$\begin{aligned} \varphi_0(t) &= t, \\ \varphi_k(t) &= \int_0^t (t - t_1) p(t_1) \varphi_{k-1}(t_1)\, dt_1, \qquad k = 1, 2, \dots; \end{aligned}$$

hence,

$$\begin{aligned} \varphi(t) &= \varphi(t, -1) \\ &= t - \int_0^t (t - t_1) p(t_1) t_1\, dt_1 \\ &\quad + \int_0^t (t - t_1) p(t_1)\, dt_1 \int_0^{t_1} (t_1 - t_2) t_2 p(t_2)\, dt_2 \\ &\quad + \cdots + (-1)^k \int_0^t (t - t_1) p(t_1)\, dt_1 \cdots \int_0^{t_{k-1}} (t_{k-1} - t_k) t_k p(t_k)\, dt_k + \cdots. \end{aligned}$$

To determine the Lyapunov constant, we need an expression for $\dot{\varphi}(t)$; we write it as follows:

$$\begin{aligned}\dot{\varphi}(t) = 1 &- \int_0^t p(t_1)t_1\,dt_1 \\ &+ \int_0^t p(t_1)\,dt_1 \int_0^{t_1} (t_1 - t_2)t_2 p(t_2)\,dt_2 \\ &+ \cdots + (-1)^k \int_0^t p(t_1)\,dt_1 \cdots \int_0^{t_{k-1}} (t_{k-1} - t_k)t_k p(t_k)\,dt_k + \cdots .\end{aligned} \tag{6.3.13}$$

From (6.3.6), (6.3.12), and (6.3.13) we obtain

$$\begin{aligned} A &= f(\omega) + \dot{\varphi}(\omega) \\ &= 2 - \omega \int_0^\omega p(t_1)\,dt_1 \\ &\quad + \int_0^\omega dt_1 \int_0^{t_1} (\omega - t_1 + t_2)(t_1 - t_2)p(t_1)p(t_2)\,dt_2 \\ &\quad + \cdots + (-1)^k \int_0^\omega dt_1 \int_0^{t_1} dt_2 \cdots \int_0^{t_{k-1}} (\omega - t_1 + t_k)(t_1 - t_2) \\ &\qquad \times (t_2 - t_3) \cdots (t_{k-1} - t_k) \\ &\qquad \times p(t_1) \cdots p(t_k)\,dt_k + \cdots .\end{aligned} \tag{6.3.14}$$

Now we turn to concrete results.

THEOREM 6.3.1 (Lyapunov [**19**]). *If*

$$p(t) \leqslant 0, \qquad p \in C(\mathbb{R}), \qquad p(t) \not\equiv 0,$$

*then equation* (6.3.1) *is unstable; moreover, both multipliers are positive, one of them being greater than one, and the other less than one.*

PROOF. The equality (6.3.14) for $p(t) \leqslant 0$ implies that $A > 2$; therefore, our statement holds. The positivity of both multipliers for $A > 2$ follows from (6.3.5). □

THEOREM 6.3.2 (Lyapunov [**19**]). *If* $p(t) \geqslant 0$, $p \in C(\mathbb{R})$, *and*

$$0 < \omega \int_0^\omega p(t)\,dt \leqslant 4, \tag{6.3.15}$$

*then equation* (6.3.1) *is stable; the multipliers are complex conjugate and lie on the unit circle.*

PROOF. Let us write (6.3.14) in the form

$$2 - A = I_1 - I_2 + I_3 - \cdots \tag{6.3.16}$$

and show that the series on the right of the formula (6.3.16) is a Leibnitz series. Indeed, this series is alternating, the absolute values of its terms decrease monotonically, and $\lim_{k\to\infty} I_k = 0$. The first property follows from (6.3.14) for $p(t) \geqslant 0$, and the third, from the convergence of the series (6.3.14). Let us show the second property. The general form of $I_k$ is given in (6.3.14). Thus,

$$I_{k+1} = \int_0^{\omega} dt_1 \cdots \int_0^{t_{k-1}} dt_k (t_1 - t_2)(t_2 - t_3) \cdots (t_{k-1} - t_k) p(t_1) \cdots p(t_k)$$
$$\times \int_0^{t_k} (\omega - t_1 + t_{k+1})(t_k - t_{k+1}) p(t_{k+1})\, dt_{k+1}.$$

Using the evident inequality $xy \leqslant (x+y)^2/4$, we find

$$(\omega - t_1 + t_{k+1})(t_k - t_{k+1}) \leqslant \frac{1}{4}(\omega - t_1 + t_k)^2 < \frac{\omega}{4}(\omega - t_1 + t_k),$$

and, finally,

$$I_{k+1} < \frac{\omega}{4} \int_0^{\omega} p(t)\, dt \cdot I_k \leqslant I_k.$$

From the last inequality it follows that $I_k$ decreases monotonically with the growth of $k$. From the estimate $0 < S < I_1$ for the sum of a Leibnitz series, in the case (6.3.16) we obtain

$$0 < 2 - A < \omega \int_0^{\omega} p(t)\, dt \leqslant 4;$$

hence, $|A| < 2$; therefore, Theorem 6.3.2 is valid. □

COROLLARY 6.3.1. *If equation* (6.3.1) *has an unbounded solution for* $p(t) \geqslant 0$, *then*

$$\omega \int_0^{\omega} p(u)\, du > 4.$$

EXAMPLE 6.3.1. For which values of the real parameters $a$ and $b > 0$ is the equation $\ddot{x} + (a + \sin bt)x = 0$ stable and for which is it unstable?

1) Let us check for which $a$ and $b > 0$ the condition (6.3.15) holds. Here $\omega = 2\pi/b$,

a) $p(t) \geqslant 0 \quad \Rightarrow \quad a \geqslant 1,$

b) $0 < \dfrac{2\pi}{b} \displaystyle\int_0^{2\pi/b} (a + \sin bt)\, dt = \dfrac{2\pi}{b} \cdot \dfrac{2\pi}{b} a \leqslant 4 \quad \Rightarrow \quad \dfrac{\pi^2 a}{b^2} \leqslant 1.$

Finally, $b^2 \geqslant \pi^2 a \geqslant \pi^2$.

2) The condition $a + \sin bt \leqslant 0$ gives the answer to the second question, i.e., $a \leqslant -1, b \in \mathbb{R}_+$.

The following result weakens the conditions of Theorem 6.3.2, taking into account, e.g., the possibility of the coefficient $p(t)$ having alternating sign. First we prove a lemma.

LEMMA 6.3.1. *Let*

$$\int_0^{\omega} p(t)\, dt \geqslant 0. \tag{6.3.17}$$

*Then the normal solution of equation* (6.3.1) *corresponding to a real multiplier has a zero on any closed interval of length* $\omega$.

PROOF. Let $\rho$ be a real multiplier of equation (6.3.1). According to Theorem 1.4.1, a nontrivial real solution $x(t)$ such that

$$x(t+\omega) = \rho x(t) \tag{6.3.18}$$

corresponds to it. Assume that $x(t) > 0$ for $t \in [0, \omega]$, contrary to the claim. It follows then from (6.3.18) that $x(t) > 0$ for $x \in \mathbb{R}$. Substitute this solution in equation (6.3.1), divide it by $x(t)$, and integrate the identity obtained from 0 to $\omega$:

$$\int_0^\omega \frac{x''(t)}{x(t)}\, dt + \int_0^\omega p(t)\, dt = 0,$$

or

$$\frac{x'(t)}{x(t)}\bigg|_0^\omega + \int_0^\omega \left(\frac{x'(t)}{x(t)}\right)^2 dt + \int_0^\omega p(\tau)\, d\tau = 0.$$

By virtue of (6.3.18), the first summand in the last identity is equal to zero, the second is positive, and we come to a contradiction with the condition (6.3.17). □

THEOREM 6.3.3. *Let*
1) $\int_0^\omega p(t)\, dt \geqslant 0$,
2) $\omega \int_0^\omega p^+(t)\, dt \leqslant 4, \quad p^+(t) \equiv \max(p(t), 0)$.
*Then equation* (6.3.1) *is stable.*

PROOF. Let us show that under the conditions of the theorem all the solutions of equation (6.3.1) are bounded; hence, by Corollary 6.2.3 they are stable.

Assume the opposite: let equation (6.3.1) have an unbounded solution. Propositions 6.3.1–6.3.3 imply that this is possible only if there exists a real multiplier. To each real multiplier $\rho$ there corresponds a normal solution $x(t)$, which, under the first condition of the theorem by virtue of Lemma 6.3.1, necessarily has a zero on the closed interval $[0, \omega]$. The identity (6.3.18) implies that the distance between consecutive zeros of $x(t)$ does not exceed $\omega$. Let

$$x(t_0) = x(t_1) = 0 \qquad \text{and} \qquad t_0, t_1 \in [0, \omega].$$

By Theorem 6.1.5, we have

$$\omega \int_0^\omega p^+(\tau)\, d\tau > 4;$$

this contradicts the second condition of the theorem. □

EXAMPLE 6.3.2. For which values of the real parameters $a$ and $b > 0$ is the equation $\ddot{x} + (a + \sin bt)x = 0$ stable? Let us check the conditions of the theorem:

$$\int_0^{2\pi/b} (a + \sin bt)\, dt = \frac{2\pi a}{b} \geqslant \quad \Rightarrow \quad a \geqslant 0.$$

If $a > 1$, then $p(t) \geqslant 0$ and the conditions of Theorem 6.3.2 are satisfied (the answer is given in Example 6.3.1).

Now let $0 \leqslant a < 1$. Then

$$\int_0^{2\pi/b} (a + \sin bt)^+\, dt = \int_0^{\pi/b + (\arcsin a)/b} (a + \sin bt)\, dt + \int_{2\pi/b - (\arcsin a)/b}^{2\pi/b} (a + \sin bt)\, dt$$
$$= (a\pi + 2a \arcsin a + 2\sqrt{1 - a^2}/b).$$

In this case we obtain

$$1 > a \geqslant 0,$$
$$2\pi(a\pi + 2a \arcsin a + 2\sqrt{1 - a^2})/b^2 \leqslant 4.$$

# Appendix

## 1. The Gronwall-Bellman lemma and its generalization.

THE GRONWALL-BELLMAN LEMMA [**19**]. *Let functions $u(t) \geqslant 0$, $v(t) \geqslant 0$ be defined and continuous for $t \geqslant t_0$ and*

$$u(t) \leqslant \lambda + \int_{t_0}^{t} u(\tau)v(\tau)\,d\tau, \tag{A.1}$$

*where $\lambda \in \mathbb{R}_+$. Then for $t \geqslant t_0$ we have*

$$u(t) \leqslant \lambda e^{\int_{t_0}^{t} v(\tau)\,d\tau}. \tag{A.2}$$

PROOF. Denote the right-hand side of the inequality (A.1) by $g(t)$; hence,

$$g'(t) = u(t)v(t) \leqslant g(t)v(t),$$

or $g'(t) - v(t)g(t) \leqslant 0$. Multiplying the last inequality by $\exp\left(-\int_{t_0}^{t} v(\tau)\,d\tau\right)$, we write

$$d\left[g(t)\exp\left(-\int_{t_0}^{t} v(\tau)\,d\tau\right)\right] \leqslant 0.$$

Integrating the result from $t_0$ to $t$, we obtain

$$g(t)\exp\left(-\int_{t_0}^{t} v(\tau)\,d\tau\right) - g(t_0) \leqslant 0,$$

or $g(t) \leqslant \lambda \exp \int_{t_0}^{t} v(\tau)\,d\tau$. Taking into account that $u(t) \leqslant g(t)$, we obtain (A.2). □

COROLLARY. *If in the conditions of the lemma $\lambda = 0$, then $u(t) \equiv 0$ for $t \geqslant t_0$.*

A GENERALIZATION OF THE GRONWALL-BELLMAN LEMMA. *Let the function $u(t)$ be positive and continuous for $t \in (a,b)$ and let it satisfy the integral inequality*

$$u(t) \leqslant u(\tau) + \left|\int_{\tau}^{t} u(z)v(z)\,dz\right| \tag{A.3}$$

*for any $t, \tau \in (a,b)$, where $v \in C(a,b)$ and $v(t) \geqslant 0$, $t \in (a,b)$. Then the two-sided estimate*

$$u(t_0)e^{-\int_{t_0}^{t} v(z)\,dx} \leqslant u(t) \leqslant u(t_0)e^{\int_{t_0}^{t} v(z)\,dz} \tag{A.4}$$

*is valid for $a < t_0 \leqslant t < b$.*

PROOF. 1) $t \geqslant \tau$. In this case the inequality (A.3) has the form

$$u(t) \leqslant u(\tau) + \int_\tau^t u(z)v(z)\,dz$$

and the right-hand side of the estimate (A.4) follows from the last inequality by the Gronwall-Bellman lemma for $\tau = t_0$.

2) $t \leqslant \tau$. In this case the inequality (A.3) can be rewritten in the following way:

$$u(t) \leqslant u(\tau) + \int_t^\tau u(z)v(z)\,dz. \tag{A.5}$$

Denoting the right-hand side of (A.5) by $g(t)$, we have

$$g'(t) = -u(t)v(t) \geqslant -v(t)g(t),$$

or

$$g'(t) + v(t)g(t) \geqslant 0.$$

Multiplying the last inequality by

$$\exp\left(\int_\tau^t v(z)\,dz\right),$$

we obtain

$$d\left(g(t)\exp\int_\tau^t v(z)\,dz\right) \geqslant 0. \tag{A.6}$$

Let us integrate (A.6) from $t$ to $\tau$; then

$$g(\tau) - g(t)\exp\int_\tau^t v(z)\,dz \geqslant 0, \qquad \text{or} \qquad g(\tau) \geqslant g(t)\exp\int_\tau^t v(z)\,dz.$$

Taking into account that $g(\tau) = u(\tau)$ and $g(t) \geqslant u(t)$, we have

$$u(\tau) \geqslant u(t)\exp\int_\tau^t v(z)\,dz = u(t)\exp\left(-\int_t^\tau v(z)\,dz\right). \tag{A.7}$$

Recall that $\tau \geqslant t$. Changing $t$ to $t_0$ and $\tau$ to $t$ in (A.7), we obtain the left-hand side of the estimate (A.4). □

**2. Regularity of a linear system almost reducible to an autonomous one.**

THEOREM 3.5.2. *A linear system almost reducible to a system with constant coefficients is regular.*

PROOF. Let a system

$$\dot{x} = A(t)x, \qquad A \in C(\mathbb{R}_+), \tag{A.8}$$

be almost reducible to a system $\dot{y} = By$, where $B$ is a constant matrix. According to Theorem 3.4.3, the system (A.8) is also almost reducible to a diagonal system, whose diagonal consists of the real parts of the eigenvalues of the matrix $B$. Denote them by $\lambda_1, \lambda_2, \dots, \lambda_n$. Thus, for any $\delta > 0$ there exists a Lyapunov transformation $x = L_\delta(t)z$ reducing the system (A.8) to the system

$$\dot{z} = \operatorname{diag}[\lambda_1, \lambda_2, \dots, \lambda_n]z + \Phi(t)z, \tag{A.9}$$

where $\sup \|\Phi(t)\| \leqslant \delta$. By the stability of characteristic exponents of systems with constant coefficients the spectrum of the system (A.9) is $\delta$-close to $\{\lambda_1, \lambda_2, \dots, \lambda_n\}$, and, by the properties of Lyapunov transformations, it coincides with the spectrum of the system (A.8).

Let $X(t)$ and $Z(t)$ be normal fundamental matrices of the systems (A.8) and (A.9) such that $X(t) = L_\delta(t)Z$; therefore,

$$\operatorname{Det} X(t) = \operatorname{Det} L_\delta(t) \operatorname{Det} Z(t).$$

Hence, by the Ostrogradskiĭ-Liouville formula, we have

$$\begin{aligned} &\operatorname{Det} X(t_0) \exp \int_{t_0}^{t} \operatorname{Sp} A(\tau)\, d\tau \\ &\quad = \operatorname{Det} L_\delta(t) \operatorname{Det} Z(t_0) \exp \left( \int_{t_0}^{t} \sum_{i=1}^{n} \lambda_i \, d\tau + \int_{t_0}^{t} \operatorname{Sp} \Phi(\tau)\, d\tau \right). \end{aligned}$$

Passing to absolute values in both sides of the last equality, taking logarithms, and dividing by $t$, we obtain

$$\begin{aligned} &\frac{1}{t} \int_{t_0}^{t} \operatorname{Re} \operatorname{Sp} A(\tau)\, d\tau \\ &\quad = \frac{1}{t} \ln |\operatorname{Det}(L_\delta(t) Z(t_0) X^{-1}(t_0))| + \frac{t - t_0}{t} \sum_{i=1}^{n} \lambda_i + \frac{1}{t} \int_{t_0}^{t} \operatorname{Re} \operatorname{Sp} \Phi(\tau)\, d\tau. \end{aligned}$$

Note that $|\operatorname{Sp} \Phi(t)| \leqslant n\delta$. Thus,

$$\begin{aligned} &\frac{t - t_0}{t} \sum_{i=1}^{n} \lambda_i + \frac{1}{t} \ln |\operatorname{Det}(L_\delta(t) Z(t_0) X^{-1}(t_0))| - n\delta \frac{t - t_0}{t} \\ &\quad \leqslant \frac{1}{t} \int_{t_0}^{t} \operatorname{Re} \operatorname{Sp} A(\tau)\, d\tau \\ &\quad \leqslant \frac{1}{t} \ln |\operatorname{Det} L_\delta(t) Z(t_0) X^{-1}(t_0)| + n\delta \frac{t - t_0}{t} + \frac{t - t_0}{t} \sum_{i=1}^{n} \lambda_i . \end{aligned}$$

In the last inequality we pass to the limit as $t \to \infty$, taking into account that $\sup_{t \in \mathbb{R}_+} |\operatorname{Det} L_\delta(t)|$ is bounded. By the squeeze convergence principle, we obtain that the limit

$$\lim_{t \to \infty} \frac{1}{t} \int_{t_0}^{t} \operatorname{Re} \operatorname{Sp} A(\tau)\, d\tau$$

exists and

$$\sum_{i=1}^{n} \lambda_i - n\delta \leqslant \lim_{t \to \infty} \frac{1}{t} \int_{t_0}^{t} \operatorname{Re} \operatorname{Sp} A(\tau)\, d\tau \leqslant \sum_{i=1}^{n} \lambda_i + n\delta.$$

Since $\delta > 0$ is arbitrary, we have

$$\lim_{t \to \infty} \frac{1}{t} \int_{t_0}^{t} \operatorname{Re} \operatorname{Sp} A(\tau)\, d\tau = \sum_{i=1}^{n} \lambda_i;$$

this, by Lemma 3.5.1, shows that system (A.8) is regular. □

# References


1. V. M. Alekseev, *On the asymptotic behavior of solutions of weakly nonlinear systems of ordinary differential equations*, Dokl. Akad. Nauk SSSR **134** (1960), no. 2, 247–250; English transl. in Soviet Math. Dokl. **1** (1960).
2. V. M. Alekseev and R. È. Vinograd, *To the method of "freezing"*, Vestnik Moskov. Univ. Ser I Mat. Mekh. (1966), no. 5, 30–35. (Russian)
3. V. P. Basov, *On the structure of a solution of a regular system*, Vestnik Leningrad. Univ. (1952), no. 12, 3–8. (Russian)
4. R. Bellman, *Stability theory of differential equations*, McGraw-Hill, New York, 1953.
5. Yu. N. Bibikov, *A general course of ordinary differential equations*, Leningrad Univ., Leningrad, 1981. (Russian)
6. Yu. S. Bogdanov, *To the theory of systems of linear differential equations*, Dokl. Akad. Nauk SSSR **104** (1955), no. 6, 813–814. (Russian)
7. ———, *A note on §8 of I. G. Malkin's monograph "The theory of stability of motion"*, Prikl. Mat. Mekh. **20** (1956), no. 3, 448. (Russian)
8. ———, *Characteristic numbers of systems of linear differential equations*, Mat. Sb. **41** (1957), no. 4, 481–498. (Russian)
9. B. F. Bylov, R. È. Vinograd, D. M. Grobman, and V. V. Nemyckiĭ, *The theory of Lyapunov exponents and its applications to problems of stability*, "Nauka", Moscow, 1966. (Russian)
10. B. F. Bylov, *Almost reducible systems of differential equations*, Sibirsk. Mat. Zh. **3** (1962), no. 3, 333–359. (Russian)
11. ———, *On the reduction of systems of linear equations to the diagonal form*, Mat. Sb. **67** (1965), no. 3, 338–344; English transl. in Amer. Math. Soc. Transl. Ser. 2 **89** (1970), 51–59.
12. B. F. Bylov and N. A. Izobov, *Necessary and sufficient conditions for stability of characteristic exponents of a diagonal system*, Differentsial′nye Uravneniya **5** (1969), no. 10, 1785–1793; English transl. in Differential Equations **5** (1969).
13. ———, *Necessary and sufficient conditions for stability of characteristic exponents of a linear system*, Differentsial′nye Uravneniya **5** (1969), no. 10, 1794–1903; English transl. in Differential Equations **5** (1969).
14. B. F. Bylov, *On the reduction of a linear system to the block-triangular form*, Differentsial′nye Uravneniya **23** (1987), no. 12, 2027–2031; English transl. in Differential Equations **23** (1987).
15. R. È. Vinograd, *A new proof of the Perron theorem and some properties of regular systems*, Uspekhi Mat. Nauk **9** (1954), no. 2, 129–136. (Russian)
16. F. R. Gantmakher, *The theory of matrices*, "Nauka", Moscow, 1967; English transl., Chelsea, New York, 1959.
17. D. M. Grobman, *Characteristic exponents of systems close to linear ones*, Mat. Sb. **30** (1952), 121–166. (Russian)
18. ———, *Systems of differential equations analogous to linear ones*, Dokl. Akad. Nauk SSSR **86** (1952), no. 1, 19–22. (Russian)
19. B. P. Demidovich, *Lectures on the mathematical theory of stability*, "Nauka", Moscow, 1967. (Russian)
20. N. P. Erugin, *Reducible systems*, Trudy Mat. Inst. Steklov. **13** (1946), 1–96. (Russian)
21. ———, *Linear systems of ordinary differential equations with periodic and quasi-periodic coefficients*, Izdat. Akad. Nauk BSSR, Minsk, 1963; English transl., Math. Sci. Engrg., vol. 28, Academic Press, New York and London, 1966.
22. L. D. Ivanov, *Variations of sets and functions*, "Nauka", Moscow, 1975. (Russian)
23. N. A. Izobov, *Linear systems of ordinary differential equations*, Itogi Nauki i Tekhniki: Mat. Anal., vol. 12, VINITI, Moscow, 1974, pp. 71–146; English transl. in J. Soviet Math. **5** (1976).

24. S. M. Lozinskiĭ, *Estimates for errors of numerical integration of ordinary differential equations.* I, Izv. Vyssh. Uchebn. Zaved. Mat. **1958**, no. 5, 52–90. (Russian)
25. A. M. Lyapunov, *General problem of the stability of motion*, Collected works, vol. 2, Izdat. Akad. Nauk SSSR, Moscow, 1956. (Russian)
26. I. G. Malkin, *Theory of stability of motion*, "Nauka", Moscow, 1966. (Russian)
27. M. Marcus and H. Minc, *A survey of matrix theory and matrix inequalities*, Allyn & Bacon, Boston, 1964.
28. V. M. Millionshchikov, *A criterion for a small change of directions of solutions of a linear system of differential equations under small perturbations of the coefficients of the system*, Mat. Zametki **4** (1968), no. 2, 173–180; English transl. in Math. Notes **4** (1968).
29. ______, *A proof of attainability of central exponents of linear systems*, Sibirsk. Mat. Zh. **10** (1969), no. 10, 99–104; English transl. in Siberian Math. J. **10** (1969).
30. ______, *On the instability of singular exponents and on the asymmetry of the relation of almost reducibility for linear systems of differential equations*, Differentsial′nye Uravneniya **5** (1969), no. 4, 749-750; English transl. in Differential Equations **5** (1969).
31. ______, *Structurally stable properties of linear systems of differential equations*, Differentsial′nye Uravneniya **5** (1969), no. 10, 1775–1784; English transl. in Differential Equations **5** (1969).
32. V. V. Nemytskiĭ and V. V. Stepanov, *Qualitative theory of differential equations*, GITTL, Moscow and Leningrad, 1949; English transl., Princeton Univ. Press, Princeton, NJ, 1966.
33. I. G. Petrovskiĭ, *Lectures on the theory of ordinary differential equations*, "Nauka", Moscow, 1965; English transl., *Ordinary differential equations*, Prentice-Hall, Englewood Cliffs, NJ, 1966.
34. W. Rudin, *Principles of mathematical analysis*, McGraw-Hill, New York, 1976.
35. F. Tricomi, *Equazioni differenziale*, Einaudi, Torino, 1965.
36. D. K. Faddeev and V. N. Faddeeva, *Computational methods of linear algebra*, Fizmatgiz, Moscow, 1963; English transl., Freeman, San Francisco, CA, 1963.
37. P. Hartman, *Ordinary differential equations*, Wiley, New York, 1964.
38. L. Cesari, *Asymptotic behavior and stability problems in ordinary differential equations*, Springer-Verlag, Berlin, 1959.
39. V. A. Yakubovich and V. M. Starzhinskiĭ, *Linear differential equations with periodic coefficients and their applications*, "Nauka", Moscow, 1972; English transl., vols. 1, 2, Israel Program for Scientific Translations, Jerusalem, and Wiley, New York, 1975.
40. ______, *Parametric resonance in linear systems*, "Nauka", Moscow, 1987. (Russian)

# Index

# Other Titles in This Series

*(Continued from the front of this publication)*



(See the AMS catalog for earlier titles)